THE
RED
ARROWS

WING COMMANDER DAVID MONTENEGRO

THE
RED
ARROWS

THE OFFICIAL STORY OF
BRITAIN'S ICONIC DISPLAY TEAM

CENTURY

1 3 5 7 9 10 8 6 4 2

Century
20 Vauxhall Bridge Road
London SW1V 2SA

Century is part of the Penguin Random Hou
of companies whose addresses can be found
penguinrandomhouse.com.

Penguin
Random House
UK

Copyright © UK Crown Copyright 2022

RAF logos are trademarks of the UK Secretary of State for Defence
and used under licence. This book contains public sector information
licensed under the Open Government Licence v3.0.

With thanks to the UK MOD for its assistance in producing this work.

First published by Century in 2022

www.penguin.co.uk

A CIP catalogue record for this book is available from the British Library.

Hardback ISBN 9781529135527
Ebook ISBN 9781529157192
Audiobook ISBN 9781473594562

Set in 11.2/15 pt Minion Pro
Typeset by Jouve (UK), Milton Keynes
Printed and bound in Great Britain by Clays Ltd, Elcograf S.p.A.

The authorised representative in the EEA is
Penguin Random House Ireland, Morrison Chambers,
32 Nassau Street, Dublin D02 YH68

For those that dream to fly

Contents

Prologue

It's the second day of training in Crete and we are focusing on getting our turns and rolls really crisp. Our jets are in good order and the sky is a bright and clear blue. Mike Ling, my synchro-pair leader, and I are feeling well rested and confident in our manoeuvres – that we have really cracked the synchro pair training programme. And yet we are about to go from everything slotting into place perfectly to almost killing each other within a second.

We are rehearsing the opposition barrel roll, a manoeuvre that requires us to cross our flight paths prior to doing the opposition pass, and we know that we have a minimum distance at which we can safely cross paths.

I make the 'threshold' radio transmission and hear Mike, flying Red 6, do the same, confirming that we are both happy to enter the manoeuvre, but as I look at the view ahead of me, I suddenly realise that something is different. The aircraft opposite – Mike's aircraft – looks much bigger and closer than it should.

In that instant, I realise that the crucial crossing of our flight paths is no longer an option. We're far too close together, and the distance is shrinking with every passing moment. I'm staring at the oncoming jet, my mind racing as I quickly try to decide what to do next. It's all coming down to a matter of seconds – split-seconds, even – the distance closing rapidly between Mike's Hawk and mine.

I swiftly think through the prescribed escape manoeuvre for just this scenario, but I have a horrible feeling that, based on our current positions, the usual escape protocol might actually put Mike and I in more danger in this instance – putting the belly of my aircraft firmly into Mike's Hawk. I need to try another method of escape, and fast.

There's barely any time to think. As our jets rush towards each other, I push forward on the control column, trying to avoid Mike by going down rather than up; to fly beneath him, instead of rolling away. I figure if I can see him the whole time, I'll be able to avoid him. In the moment, I abandon the conditioned response: the escape manoeuvre that Mike and I had drilled over and over with the rest of the team. I'm flying by instinct now.

But so is Mike and, as I see him push down on his control panel, my stomach lurches as I realise we have unfortunately made the very same decision.

My heart is hammering and Mike's aircraft is still hurtling closer and closer to mine. With a pang, I understand that we are going to collide. There's no get-out here. I'm done. I duck my head and shut my eyes – I really don't want to see what is going to happen next – and wait for the inevitable.

I hear a deep thud, and my Hawk's airframe vibrates and shudders momentarily.

I'm still here. I open my eyes and all I can see is the runway hurtling towards me as my Hawk heads straight down towards it. We'd not been more than 110 feet up, and now I'm riding my jet straight into the ground. Fuelled with adrenaline, and with every sinew of my body on red alert, I pull back on the control column and my aircraft lurches upwards back into the sky.

As I fly up and away from the scene, I'm in state of total shock. Then I see an explosion in my mirror – Mike's jet hitting the ground. His Hawk splinters on impact, littering its burning parts across the airfield below.

I try and slow down my breathing, and attempt to calm myself as I look for a safe place to land my damaged aircraft. Waves of guilt are threatening to overwhelm me, but I'm not safe yet. I can't lose focus.

Not now.

1

Welcome

W E   W E R E   H E A D E D to Pasto, a Colombian city in the foot-
hills of the Andes cordillera and the family home of my
father. Our mode of transport: a rather rickety-looking propeller
aeroplane, almost pocket-sized compared to the vast commercial
airliner that we had left at Bogotá. As we climbed over the mountains of Colombia the aircraft lurched and creaked, and I remember my mother gripping my hand, petrified that we weren't going
to make it. I must have been all of eight years old at the time but,
instead of feeling anxious, I found myself enthralled by the reeling
sensations of the flight – it was simply brilliant.

After that, it was all about aviation. Air Avianca, the national
airline of Colombia, which whisked me from England via Vene-
zuela to visit my father's family in Bogotá, seemed to me as a child
the epitome of glamour. The smell of aviation fuel, getting onto the
aircraft, the experience of being on board – I couldn't get enough
of it and, as I got older, my passion for flying grew with me. At the
age of twelve I was taken to my first air show at the former RAF
station of Biggin Hill. Sitting in the grandstand with my family
on a hot summer's day, I took in everything that a good air show

delivered during that era, and I revelled in the immense power of the Tornado, the agility of the Harrier in hover and the sleek line of the SEPECAT Jaguar, not to mention the mesmerising display of the Red Arrows, who had been thrilling crowds at Biggin Hill ever since their first public UK display on 15 May 1965. The noise of those jet engines, my body vibrating with the afterburners, was overwhelming and my schoolboy brain was fizzing with 'This is awesome – imagine flying *that*!' In that instant I knew I wanted to be a pilot, and I continued to feed my obsession with further visits to Biggin Hill and various air shows around the UK.

The first time I managed to get close to flying an aeroplane was some two years later, this time with the air cadets at RAF Manston where, in the skies over Kent, my instructor took me through some loops and rolls in a two-seater Chipmunk. Little did I know then that I would return to Manston many years later as a Red Arrows team pilot, landing and taking off in a distinctive Hawk T1 jet. In the years in between, I launched my flying career as an officer cadet at the Manchester and Salford University Air Squadron, before joining the Royal Air Force (RAF) in 1999. This was followed by officer training at RAF Cranwell and then selection as a fast-jet pilot – something I had yearned for ever since those early Biggin Hill days. Advanced jet training and then selection as an instructor at RAF Valley led to front-line tours on the Tornado F3, all of which enabled me to meet the minimum criteria demanded of a prospective Red Arrows pilot, joining the team for the 2009 season.

Throughout it all – as a Red Arrows team pilot, as Red 1 team leader and now as Officer Commanding (OC) of the Red Arrows or, to give them their official title, the Royal Air Force Aerobatic Team (RAFAT) – my love of flying has never dulled. It is a passion that unites all the 160 display pilots who have flown with the team since its formation in 1964. Some of those who have worn the famous red flying suit already had the Red Arrows on their radar as they steered their RAF careers towards joining the team, while others simply wanted to fly fighter jets on the front line and found a natural home within the Red Arrows.

Many team pilots have family members in the armed forces, and their fathers or grandfathers flew fighters or bombers in the Second World War or in more recent conflicts, such as former team pilot Kirsty Murphy (née Stewart), whose father served as a navigator in the Gulf War. The longest-serving team pilot, Mike Ling, set his sights on joining the Red Arrows after seeing the team at Biggin Hill when he was very small. And recent Red 1 Martin Pert remembers the thrill of seeing the team when he was seven or eight at the annual Rolls-Royce Air Show in Hertfordshire, his house being so close to the display site that the famous red, white and blue vapour trails would hang in the air over his parents' back garden. However, some Red Arrows pilots are drawn to the team from much further afield: the legendary Ray Hanna, who led the team in the 1960s, hailed from New Zealand. So, too, did former Red 1 Ian Dick, who remembers marvelling as a young RAF cadet at the formation aerobatics that he saw at Farnborough Air Show, which far surpassed anything he'd seen back home in New Zealand.

On joining the Red Arrows, that passion for flying becomes inevitably mixed with immense pride in being part of one of the best aerobatic display teams in the world – one that represents the excellence and incredible capabilities of the Royal Air Force – while also acting as an ambassador for the United Kingdom, both at home and abroad. As Officer Commanding, I'm honoured to head up the 125-strong team at the Red Arrows, made up of nine display pilots – all of them fast-jet pilots from front-line RAF squadrons – as well as around eighty-five highly trained engineers, technicians and support personnel, known as the Blues (after the blue suits they wear). Ten of those engineers and a photographer – a group known as the Circus – are chosen to fly in the passenger seats of the team's Hawks during the display season, an unusual and much sought-after role in the RAF.

Since they were formed in 1964 the Red Arrows have been thrilling crowds with precision flying and perfectly symmetrical formations, their nine Hawk T1 jets just a few feet apart. Over its fifty-eight years, the team has performed almost 5,000 times across the globe, in displays that are filled with aerobatic feats and knife-edge passes, with the jets travelling at speeds of 350–400mph. Each year the team performs at multiple air shows and festivals, which have a long history of attracting huge crowds, as they still do today: some 4.9 million people attended air shows in the UK in 2019. The Red Arrows also perform at more than a hundred fly-pasts each year, from those that mark major national events, such as VE Day celebrations, the Queen's Birthday Fly-past and the Great North Run, to smaller local events across the four nations of the United Kingdom. This means that some ten million people see the Red Arrows every year – and that doesn't factor in the many appearances the team makes in mainland Europe, the Middle East and around the globe. A familiar and magnificent sight in the skies, the now-famous Diamond Nine formation represents the peak of precision flying and excellence, a measure by which other high-performance teams compare themselves. If a sporting team is working together in a seamless and synchronised way, whether it's on a football or rugby pitch, a racing circuit or in any other arena, then a pundit may well compare it to the Red Arrows – a name that has transcended the world of aviation and has passed into cultural heritage, its nine jets symbolising the pinnacle of high-performance teamwork.

My own journey with the Royal Air Force Aerobatic Team began in 2008 when I was selected for the team. I immediately began the rigorous training programme, and gradually built up the flying skills and mental resilience required to achieve the benchmark of excellence – or *Éclat*, as the Red Arrows' motto states. I was already an experienced fast-jet pilot, having flown Tornados on the front line and having instructed in the Hawk, but the Red Arrows require a particular form of precision flying that can only be learned through repeated flights, or 'sorties', over six months, gradually building up the formation and finessing moves, with safety being paramount throughout.

In my first year with the Red Arrows I flew as 'Red 3'. All the pilots in the formation are numbered from one to nine, with the team leader, known as Red 1, at the front. Red 2 and Red 3 are positioned tucked next to the leader, with Red 4 and Red 5 on the outermost points of the formation. Reds 6 and 7, known as the synchro pair, make up the rear section of the formation, along with Reds 8 and 9. Aircraft on the right of Red 1 are always even numbers, while odd numbers fly on the left. The pilots' formation positions are rotated every year, with pilots often flying in more challenging positions in the second and final years of their three-year-long tour of duty. In my second year with the team I took up the Red 7 position in the synchro pair, which, in the second half of displays, flies the more dynamic and daring head-to-head passes at a combined speed of 800mph.

Being part of the synchro pair is one of the more challenging and demanding positions within the team. In early 2010, during our spring training in Crete, I experienced a near-fatal mid-air collision with my synchro-pair leader Mike Ling, leading him to eject from his jet and suffer multiple injuries, whereas I escaped with only a damaged tailfin and was able to land safely at Heraklion airport. While Mike was out recovering for the season, I was fortunate to be physically unharmed and able to carry on with a new synchro partner and complete that summer's campaign. In 2011 I became the leading half of the synchro pair, flying as Red 6, and finished my first stint with the Red Arrows in a year that, tragically, saw two fatalities. These were the first fatalities in twenty years, and both had a profound effect on the team.

After a tour in Afghanistan, I returned to the Red Arrows in 2014, this time leading the formation as Red 1 – a role that comes with immense responsibility and unrelenting pressure. In my second year as Red 1, I led the team on an Asia-Pacific and Middle Eastern tour, during which we displayed in China for the first time in Red Arrows history. Taking twelve single-engine jet aircraft across 22,000 miles and seventeen countries was not without considerable challenges – not least getting through monsoonal thunderstorms over India and Pakistan – but the tour ultimately proved a major success, reaching more than one billion people through various media and engagement activities and helping to support UK trade objectives. The year of 2017 would see another Middle Eastern tour and the usual busy schedule of displays and fly-pasts in both the UK and Europe, after which the three years of my second stint with the team were up and I left to work at the Typhoon headquarters at RAF Coningsby. In 2020 I was honoured to return to the team as Wing Commander and am now privileged to lead the world-class Royal Air Force Aerobatic Team, which, in addition to thrilling and delighting crowds in displays around the world, helps to deliver 'soft-power' capability in support of national interests both at home and overseas.

The 'soft power' of the Red Arrows has long been recognised by the UK government and British industry as a valuable national asset – not only in enthralling the millions of spectators who see our displays, but also in the vital role that the team plays in fostering alliances and helping to promote the UK abroad. The Defence Purpose of the UK's three armed forces, as defined by the Ministry of Defence (MOD), is to deliver the three National Security Objectives: to protect the people, to project global influence and to promote prosperity. At the time of writing, the Red Arrows have performed in fifty-seven countries around the globe. The team's reputation for excellence on the ground and in the air has always been widely admired by air forces and spectators across the world, and the Red Arrows are a vital component in projecting the UK's global influence. We contribute to its prosperity agenda by increasing opportunities for UK industry, by promoting employment and skills, and by emphasising a sense of pride in our country. In more recent years the Red Arrows have worked closely with the government on its 'GREAT Britain' campaign, which aims to promote the best of the country abroad, encouraging people to visit, do business, invest and study in the country, and inspire the younger generation with a focus on the STEM subjects of science, technology, engineering and maths.

In this book, the first official story of the Red Arrows, it is my aim to provide a new and unique perspective on the team. I want to share with you insights from some of the pilots who flew with the Red Arrows during its earliest days, including those who flew alongside the team's legendary founder members, Flight Lieutenant (Flt Lt) Lee Jones and Squadron Leader Ray Hanna, who performed daring aerobatic feats in single-engine Folland Gnat jets. Many of those pilots had honed their skills as fighter pilots and as team members of the Royal Air Force's various aerobatic teams, out of which the Red Arrows were born. Along with background on early air shows and the development of formation and aerobatic flying in the UK, I will take you on a journey through the key highlights of the Red Arrows team over its past fifty-eight years.

I will also go into detail on how pilots are selected for a team that demands not only exemplary fast-jet flying skills, but also a certain type of person who is able to critique themselves and their peers in a unique and high-pressure environment that depends entirely on trust. In this book I will also reveal what it is like to train with the Red Arrows: how pilots develop the unique 'fly by ear' skills needed to fly in close formation, as well as the mental resilience required to perform loops, rolls, complex manoeuvres and high-speed head-to-head passes. I'll outline the incredibly precise minute-by-minute Red Arrows 'check-in' – famous throughout the RAF – which ensures that the nine gleaming jets arrive exactly where and when they should, to the absolute second, whether that is over Wembley Stadium or the seafront at Southport or halfway around the world.

I'll also do my best to describe what it feels like to lead this remarkable team, with eight Hawks trained on your every move and radio call, as you make decisions mid-air on the types of displays to perform, or whether to abandon transits or displays altogether, in the face of bad weather. I have been fortunate to draw on the memories of previous team leaders, from Squadron Leader Ian Dick and Dickie Duckett, who first flew with the team in 1968, through to Brian Hoskins, John Rands, Jim Turner and recent team leader Martin Pert. Each provides his own insights into the challenges of being Red 1, as well as describing the many highs and lows during their combined time with the team, spanning more than six decades. My former synchro partner Mike Ling, and current Red 10 Adam Collins – the team's safety supervisor who watches displays from the ground and provides feedback to the rest of the team – both also provide a fascinating look at how some of the airborne images of the team are captured. And I speak to former fighter pilot Kirsty Murphy about her unique experience as the Red Arrows' first female team pilot.

While the team brings joy to millions, the Red Arrows have also faced dark days, when cuts in defence funding have led to questions about the future of the unit, or when accidents have highlighted the immense risks involved in flying fast jets so close together. We take our safety record very seriously, and the book examines how we mitigate all the risks associated with the delivery of formation aerobatic flying. Mike Ling and I also provide a first-hand account of our 2010 near-fatal mid-air collision, as well as insights into other incidents that have occurred over the years. A thirst for constant improvement and development is a cornerstone of the team, and we always strive to learn and build from every experience, whether on the ground or in the air.

Ben Ireland, who was Junior Engineering Officer between late 2019 and mid-2021, describes life in the Blues engineering, technical and support team, without whom the Red Arrows would be unable to function. Representing many of the trades and skills seen right across the Royal Air Force, the Blues, through hard work and dedication, work around the clock to ensure that our Hawks – some of which are more than forty years old and have little in the way of computers on board – are in peak condition and able to meet the high-octane demands the pilots place upon them.

I also talk about the many long-haul tours the Red Arrows have achieved around the world: the team's first crossing of the Atlantic, in single-engine aircraft with limited fuel capacity and rudimentary navigational aids; the challenges of flying in bad weather, including John Rands' world tour; displaying for the first time in Zimbabwe and South Africa, and in Sydney Harbour to a record crowd of 1.2 million; leading the team on groundbreaking tours to China and the Asia-Pacific region; and the most recent long-haul tour to North America in 2019 under the leadership of former Red 1, Martin Pert. The book also covers some of the more unique fly-pasts that the team has done over the years, from flying on the wing of a prototype Concorde in 1972, to an appearance with the last remaining Vulcan bomber in 2015; and the RAF's centenary fly-past in 2018, made up of a hundred RAF aircraft, one for every year.

I'll provide an insight into the extensive ground and media engagements that the team undertakes, including meeting members of the Royal Family and dignitaries around the world – not to mention the raft of elite sportsmen and women and media personalities who have visited and flown with the Red Arrows over the years. And, finally, those very special meetings with our fans and the general public: from schoolchildren in Beijing and Second World War veterans at Farnborough, to families on a breezy British seafront, and the thousands who joined our online events held throughout 2020.

It is my hope that this book will give you a fascinating and unprecedented look at what is a unique and iconic Royal Air Force unit, shedding light on the Red Arrows' core values and principles of leadership, teamwork and risk management, plus its relentless pursuit of perfection, many of which qualities underpin the ethos of the RAF and the UK armed forces as a whole. Above all, I want to convey the excitement, dedication and vibrancy of a Red Arrows show and celebrate an incredible team, made up of 125-plus people who work together to produce something quite extraordinary, year after year – something that captures the imagination of millions of people right across the world.

2

Making the Team

ICAN RECALL EVERY MINUTE of my Red Arrows flying test – it is scorched into my memory. It is conducted using two Hawk jets: Red 1 is in the lead aircraft and the prospective team pilot is in the second jet with the team's Executive Officer (XO), normally the experienced Red 8 or 9. For a seasoned RAF fighter pilot, the test might seem easy on paper: fly two formation loops and barrel rolls, change formation position, then do it again. And yet it's incredibly stressful. Prior to getting strapped in, the XO gives you a full brief on what is expected during the sortie and, once airborne, he or she says very little until you hear those fateful words: 'You have control.' That's it – you're now flying on the wing of Red 1, and the next eight minutes will probably determine whether or not you will become a Red Arrow.

While the flying test is of course critical, it is only one part of a rigorous selection process, one that is normally – bar global pandemics or fuel crises – carried out on an annual basis. The standard maximum tenure for a Red Arrows pilot is three years, and thus new recruits are required on a rolling basis, all of them fast-jet pilots drawn from Royal Air Force ranks. Only those who meet certain criteria can apply for the job: they must be assessed as an 'above average' fast-jet pilot (or better) by their superiors, have approximately 1,500 flying hours under their belt and have completed at least one front-line fast-jet tour. Such requirements mean that those who apply are usually Flight Lieutenants in their late twenties or early thirties, and applicants – many of whom don't get in first time around – can only apply a maximum of three times.

In the early days of the team, selection was not quite so formal, although all those who flew for the Red Arrows were required to be fast-jet instructors and among the best pilots the RAF had to offer. From the outset, the Red Arrows were keen to have volunteers who really wanted to be on the team, rather than being 'pressed' into joining by the upper chain of command. Not that this was ever a problem, because a position on the team has always been a sought-after role in the RAF, and the Red Arrows normally recruit only two or three new pilots each year. When the team was formed in 1964, many of those who flew with the Red Arrows were known to the original team members as exceptional pilots: the team's first leader, Flt Lt Lee Jones, brought in pilots he knew from other RAF aerobatic teams or had trained himself. Flt Lt Dickie Duckett was one of those trainee pilots, flying loops and rolls with Lee Jones at RAF Valley back in 1963, at a time when Jones was already thinking of setting up the Red Arrows. The team's second leader, Ray Hanna, would eventually call him in 1967 and invite Duckett to join the Reds.

When pilots join the RAF, many simply want to fly anything – be it multi-engine aircraft, helicopters or, in the case of those who go on to fly with the Red Arrows, fast jets. With tours that last three years, the RAF specialises in throwing pilots out of their comfort zone as they grapple with the next task, whether that's flying on the front line or acting in an administrative or staff role. Many who apply to RAFAT are looking for that next challenge, or are simply intrigued to know whether they might be skilled enough even to be considered for the Red Arrows. Such was the case with John Rands, team leader from 1994 to 1996, who had accrued several years' experience of flying jets on the front line: 'The team is a visible part of the RAF, it's been there a long time and part of you wants to know if you can do it. When you've been on a front-line squadron for a while, you're looking beyond that. Flying fighters in the Air Force is pretty cool, but the Red Arrows is the cherry on the cake.'

Martin Pert, Red 1 from 2018 to 2020, had flown the Harrier, one of the most complex aircraft in the RAF, then became a fast-jet instructor, and it was only at that point that he considered the Red Arrows as a next possible step in his career. Jim Turner, a former Jaguar pilot, weapons instructor and Red Arrows team leader from 2012 to 2014, had never considered joining the team until he was persuaded to fly with them for a day, after which he needed no further convincing, feeling that it looked like 'a great flying job and very exciting'. He quickly made it known that he was keen to join the team.

APPLYING YOURSELF

Making the decision to apply is one thing, but to be shortlisted and then get through the selection process is something else entirely, and many pilots are only successful at their second or third attempt. As a result, some first-time applicants view the process as a means of getting their name known to the squadron, and an opportunity to find out what life is really like in the Red Arrows – a team that, despite its iconic status, still retains a certain mystique. Kirsty Murphy, former Tornado fighter pilot and instructor, had few expectations of getting into the team when she first applied in 2009: 'I saw selection week as a way of setting my stall out and learning as much as I could, and I think that took off a lot of pressure.' Mike Ling, the longest-serving Red Arrow, applied early on in his Tornado career, having served eighteen months on the front line, and consequently didn't expect to get through the first time, or even to be shortlisted. However, he (like Kirsty) made it through to the team on his first attempt.

My own path to becoming a Red Arrow was similar to that of many in the team. When I joined the RAF in 1999 I was keen to fly anything, but I was extremely happy to make the grade and be selected for the fast-jet training stream. My first real steps in flight training began at the age of eighteen on securing an RAF flying scholarship at Manchester University, where I completed elementary training at the Manchester and Salford University Air Squadron – largely thanks to my unfailing and inspirational instructor Squadron Leader Rod Newman, who kept me on track when I got a little too swept up in university life (studying politics and theology) to the detriment of flight training at the weekend. On graduating, I joined the RAF, went on to officer training at RAF Cranwell, was selected as a pilot and thereafter went on to do basic fast-jet training at RAF Linton-on-Ouse, followed by advanced jet training on Hawks at RAF Valley.

I really enjoyed fast-jet training. It's a long process with lots of hurdles to clear, but as I progressed through the various phases of training, my confidence grew and I felt incredibly excited about earning the coveted RAF Wings. I was fortunate as my hard work paid off and I was delighted to graduate with the course trophy, and I assumed that I'd soon be off to the front line with my peers. Instead I was selected to stay on and train other pilots as a Qualified Flying Instructor (QFI) – or 'creamie', as it's known in RAF parlance, from the 'cream of the cream'. Although it was meant as a compliment on my skills as a pilot, this came as a bitter blow at the time, as I was focused purely on getting to my first front-line squadron, and not a single bone in my body wanted to teach.

The experience, however, would prove invaluable, as I attempted to coach and mentor a vast array of pilots, from strug-gling recruits who were not much younger than me, right through to senior officers and seasoned Gulf War veterans who had been-there, done-that, but needed a bit of a refresher course in a fast jet. I may not have realised it then, but the skills I learned as an instructor – dealing with the vagaries of human behaviour and the way they play out in the briefing room and in the cockpit – helped me not only get into the Red Arrows, but also cope with life as a team pilot, then as Red Arrows team leader, and throughout the many and varied roles that I've been lucky enough to experience in my RAF career.

As I learned during my advanced training at RAF Valley, when you're handling a fast jet like the Hawk T1, or more complex aircraft like the Typhoon, training involves not only the handling of the jet, but also a crucial conceptual element: getting your head around the fact that you're travelling on average at seven nautical miles every minute, and at times much faster – get it wrong and you can end up in a country you weren't intending to be in. Fundamental to fast-jet training is the ability to organise and structure your mind to think ahead, which is known as being 'ahead of the jet' in RAF-speak. It's not about making lots of decisions at speed – good fast-jet pilots will prioritise their thoughts in order to simplify their environment, so that the brain can make really good decisions at pace. This is a universally important skill-set, whether you are flying at twice the speed of sound in a Typhoon jet over the North Sea or leading nine Red Arrows across the UK to arrive at a display location exactly on heading, on height, at the right speed and at the right time, to the second.

After completing my tactical training, learning the basics of weapons delivery and air-combat training with eight partner nations at the NATO Flying Training School in Canada, I was selected to fly the fighter variant of the Tornado aircraft, known as the F3. A conversion course led to 'Quick Reaction Alert' (QRA) duties based at RAF Leuchars in Scotland and RAF Leeming in Yorkshire, scrambling aircraft at very short notice to intercept and identify hostile aircraft that might have strayed into UK airspace. With jets that were 'cocked' and ready to go as soon as we received the order, my job was to get airborne within moments, before flying towards my mission, which might involve anything from identifying and shepherding Russian bombers in the north, to meeting a stray civilian aircraft on UK borders that might not be responding to air-traffic control. Although flying faster and higher than Spitfires and Hurricanes, modern fighter jets of the RAF still play a critical role in defending the skies of the UK, long after the Battle of Britain in the Second World War.

Despite these front-line years on the Tornado, I had still never considered joining the Red Arrows, wrongly perceiving them as an elite few, well beyond my grasp – until my squadron boss suggested that I consider applying, perhaps seeing something in me that I hadn't detected myself. I decided to give it a go, even though it was still fairly early on in my front-line career. Usually around twenty-five to thirty RAF pilots meet the criteria and apply to join the Red Arrows each year, and I was fairly stunned to discover that I'd made it to the shortlist of nine, meaning that I was to spend six to seven days of selection with the team during their final stage of pre-season training exercises abroad (known as 'Springhawk'). As I headed to Cyprus, I simply counted myself lucky to be there and didn't really think at the time that I would make it through.

THE SHORTLIST

Pilots who apply to the Red Arrows are required to send their F5000s, records that include every flying report from day one of their training. It's Red 1's job to sift through all of these in advance, to get a sense of each candidate's raw flying ability. From these, he or she can spot negative or positive trends, get a feel for an applicant's learning ability and a flavour of the individual through their five to ten years of operational service or their time as an instructor. Red 1 then grades the applicants accordingly, based on his or her perception of their flying ability, and then presents the list to the other pilots – Reds 2 to 11. As some of the candidates will already be known to the team, the selection process is handled as sensitively and objectively as possible, with candidates being numbered and not named. The team then, as a group, grades the pilots, whittling them down to the As and B-pluses to make up the selected candidates.

The shortlisted nine are then invited to spend a week at the team's pre-season training phase. Exercise Springhawk usually takes places either at RAF Akrotiri in Cyprus or, more recently, at Tanagra Air Base, north of Athens in Greece. With blue skies and reliable weather, the team is able to train intensively, on three sorties a day, over both water and mountains – safely exposing it to the challenges of display flying over coastal sites and undulating terrain. Away from the domestic scenario, the team can get a really good look at the candidates, and it's also a chance for the shortlisted pilots to see the whole Red Arrows team in one of its most demanding training periods.

During the week the applicants fly in the back seat of every training sortie and are rotated through Reds 1 to 9. They are not in control of the aircraft, but are there to absorb the full Red Arrows experience as the team runs through all the sequences that make up the public displays, including the three types of shows we do: full, rolling and flat (see Glossary). At some point during the week, each applicant is then taken for their flying test, with everyone hoping that theirs will be early on, so that the thought of it doesn't have to loom over them for most of the day – it's such a relief to get it done.

The flying test that prospective pilots must undertake was introduced in 1992 by Wing Commander Adrian Thurley (retired), Red Arrows team leader from 1991 to 1993. The test design exists virtually unchanged today and continues to provide an objective assessment of the most suitable RAF pilots – those who are best able to cope with the intensity and demands of the first training season with the team. Prior to 1992 the selection of pilots was based on flying reports, the F5000s, along with recommendations from previous commanders. Adrian (or Ade, as he was often called) was bemused as to why there wasn't a formal flying test, considering that – although it was rare – a few pilots hadn't met the grade and needed to be replaced during the training period, or even as late as mid-season. After conducting a straw poll among ex-team leaders, Ade decided that a single flying test would offer an extra metric upon which selection could be made, specifically to identify the raw flying abilities of the candidates.

Successful formation aerobatics are impervious to the tremendous technological advancements of modern-day fighter aircraft and the vast array of skills required to operate them, and will always demand the highest degree of basic piloting skills, which are rooted in good hand–eye coordination. As a result, Ade designed a basic but very effective test: following Red 1, the candidate is instructed to fly two loops and two barrel rolls on a formation reference called Battle formation (the same reference is used when the team conducts a high-profile fly-past and creates a symmetrical V-shape). Once this is complete, they must move on to the Diamond Nine formation and do the same again. The applicants are given their formation references on the ground, prior to getting airborne, and will attempt to hold these positions during the test and while they manoeuvre the aircraft.

In conducting the test, Red 1 and the XO are not expecting perfection, and pilots will often move around quite a bit in what we call the 'safety box', in which they are never too far from or too close to the aircraft next to them. If candidates deviate from the reference, it's useful for us to see how they get themselves back into the correct position, as this demonstrates a quick learning curve. Similarly, any improvement in loops or rolls during the test can tell us a lot about the character of that pilot. During the assessment, some pilots manoeuvre the Hawk in absolute silence, while others might chat away, verbalising what they're seeing or what they need to do – 'Right, I need a bit more power', or 'I need A or B' – or even chastising themselves for not being in quite the position they want to be. I probably fall into the more verbal category, and for the XO it is of course easier to assess applicants if they are more vocal, as we can hear or see that self-awareness. I, like many other candidates, had been an instructor, so I was used to dynamically critiquing flying in the moment, as then-prospective team pilot Kirsty Murphy did during her Red Arrows flying test: 'I picked up on three errors, vocalised what they were and tried not to do them again, and it's always better to be specific about errors, rather than "I just need to be better".'

Back on the ground, the XO might ask the pilot how they think the test went, to see if they've got a good sense of what went well or not so well, which again feeds into the assessment. Pilots often think that the slightest jerk or error during the test means they have blown their chances of getting on the team. They're not given any feedback during or after the test, so they can only guess at how they were assessed. We do tell the pilots prior to the test that if control is taken from them for safety reasons it's not an automatic fail, because the way a pilot responds to that kind of situation can tell us a lot about them. To this day I still don't know what my formal grading for my test was. It felt like it went well; I felt that my second loops and rolls were better than my first, and so I must have showed some sort of learning curve. The XO, Greg Perilleux (Red 8 in 2008), didn't take control from me, so that was a bonus; and I knew that, regardless of the result, I had fulfilled a boyhood dream: I'd just flown four loops and rolls with smoke on, in a red jet being led by Red 1 – definitely a story for the grandkids. Most importantly, whatever Greg and the boss (Red 1) thought about my efforts that day, they were good enough.

With the first part over, the second element of the selection process is the formal interview, conducted by the Squadron Leader (Red 1), the Wing Commander (Officer Commanding the Red Arrows – my current role) and the Group Captain (Commander of the Hawk Headquarters that parents RAFAT). For many of the candidates, this is the first time they have been in an interview situation since they applied to join the RAF many years before. Having sat on the other side of the desk, it's amazing to see seasoned fighter pilots who have led operational missions under extreme pressure sit in an interview room in their smartest uniform in a total sweat and turning bright red, just because you've asked them a simple question.

We ask a variety of questions, probing how the pilots think their career has gone to date, where they see themselves in the future, plus some basic history-related questions about the team. When the candidates arrive on the selection week they are tasked with finding out a single fact about the Red Arrows that Red 1, the Officer Commanding or the Group Captain is unlikely to know. The annual debate for the candidates is whether to create a comical yet entirely factually incorrect answer, or whether to find an extremely arbitrary, more mundane fact and hope the panel doesn't know it. One (ultimately successful) candidate went for the latter option and, to our amusement, gave a potted history of an American truck-haulage firm, also named the Red Arrows. For those readers with aspirations to join the team, I can tell you the least best-kept secret of the interview process, the final question: 'Why do you think you should be chosen to be a Red Arrows pilot?' Even this simple question has been known to throw up some remarkable answers.

During the week, shortlisted candidates often talk amongst themselves about the kind of questions they might face, as happened before Kirsty Murphy's interview. 'There was a real debate over a couple of questions and how you should answer them,' she remembers. 'Hotly discussed were the questions "If you got into the team, who on the shortlist would you want to be on the team with?" and "If you didn't get in, who on the shortlist do you think should go on the team?" The consensus was that you shouldn't name anyone, as the Reds are very much a team, and so you shouldn't pick one person out. I thought otherwise and figured that the examiners were looking to see who was popular amongst their own cohort, so I picked someone, and it was actually the guy who was selected for the team at the same time as me.'

At the tail-end of my selection week I sat in the holding room waiting for my interview, a little nervous, although – much like the flying test – I found the whole experience incredibly novel and exciting. My subconscious had lowered any expectation of getting into the team on my first attempt, so I approached the interview as a bit of a practice run. Looking back, I think this unknowingly took some self-imposed pressure off my shoulders, and I'm sure it helped me to relax (though not too much!) in the interview and throughout the whole week.

During my interview I was asked all the usual questions, as well as a slightly more left-field one: 'Choose three people, dead or alive, whom you would invite to a private dinner party.' Before my brain had processed a well-constructed answer, I blurted out, 'Barack Obama'. By no means was I trying to make a political state-ment but, by chance, I had just finished reading *The Audacity of Hope* by the now-former US President and his name simply came out. I distinctly remember this answer grabbing the attention of the panel, and the Group Captain put his pencil down and stared at me intently. The hamster wheel in my head was now rotating at maximum revolutions per minute, in an attempt to find a suitable dinner companion for Barack Obama. Much the same mental pro-cess unfurled, with my brain reaching for previous books I'd read that year, which happened to focus on Julius Caesar in the *Emperor* series by Conn Iggulden. I meant to state Julius Caesar assertively as my second guest, but for some reason I said Cleopatra, which took me rather more by surprise than the panel. With all three senior officers now staring at me, no doubt wondering where I was going with this, after what felt like an eternity I selected my third guest, Sir Winston Churchill. Much like the flying test, I don't know what conclusions were drawn from my interview perform-ance, or indeed from my dinner-party guest selection; but, on reflection, I've often wondered how awesome it would be to host Barack Obama, Cleopatra and Sir Winston Churchill around the same dinner table.

The final formal part of the selection process is the mock media interview, which the team devised about ten years ago, motivated by the fact that Red Arrows pilots are very much the public face of the RAF and are required to interact with the media, and with large audiences, on an increasingly regular basis. Before making their selection, the team needs to see how candidates respond to questions on a range of issues, and how they cope with the odd curveball, testing how quickly they can think on their feet. Our audiences are generally welcoming and enthusiastic, but we need to be prepared for all types of scenarios. And so, in the presence of our Communications Director Andrew Morton, the selection panel asks candidates a range of quite challenging questions, which might relate to current affairs and politics, the environment, a candidate's operational history or ethical questions on military issues.

360-DEGREE INSIGHT

A less formal, but no less crucial, element of the selection process is the peer-group assessment of the candidates – opinions being canvassed from the team pilots, as well as members of the whole squadron. A variety of events are organised for the Red Arrows hopefuls, during which the candidates can relax and socialise with the team at anything from go-karting (yes, put a bunch of fast-jet pilots in any vehicle and it will get competitive), to a reception at the British Embassy in Athens. Red Arrows pilots need to be socially competent and are often fairly outgoing: meeting the public and officials at ground engagements is an important part of the job, as team leader Ray Hanna drummed into his team in the 1960s. Ian Dick similarly reminded his pilots in the early 1970s that team members shouldn't just shoot the breeze amongst themselves at events, but should instead mingle throughout the crowd. The more informal events that are put on give us extra insight into the candidates, and the candidates themselves get a good sense of what the Red Arrows are about and of the different roles of the team.

The most abiding memory I have of the selection week (as well as one of the most daunting) was when the candidates were taken to play a round of golf with the team. I had never played golf before, so I didn't relish the prospect of the team watching my feeble attempts at the game. On arriving to get on the RAF air-transport aircraft to Cyprus, I could see that the eight other candidates had their own clubs and golf bags, whereas I had borrowed my clubs from a friend who didn't have a golf bag, so I was keeping mine together in a bin-liner. As I put my bin-liner on the carousel to go into the aircraft hold, I remember thinking, 'This is so embarrassing.' Then, in what turned out to be the most nerve-racking moment of the entire week (more so than the flying test, the formal interview, and so on), I was watched by the entire Red Arrows team in their pristine uniforms as I attempted to hit the ball off the first tee. As a kid I had played a lot of cricket, so I thought to myself, 'Well, if I can hit a ball travelling towards me at speed, then surely I can hit this stationary ball.' To my immense relief, the ball went straight and quite far – and of all the adequate and, mostly, dreadful golf shots I've played since, this first tee-off is by far my most treasured memory!

At the end of the week the team's senior officers and leaders look more closely at the results from the formal assessments and collate opinions from all corners of the squadron, sifting through everything to determine which of the candidates will be ranked in the top three. It's important to hear what the team pilots think, along with the engineering and support team – the Red Arrows team works so closely together throughout the year, and trust, hard work and dedication bind everyone in it, so it's vital that they all have an input. Now, as Wing Commander, I'll make time for a good chat with the Senior Engineering Officer and the Junior Engineering Officer in the squadron, as they will have collated all the opinions of our line engineers, who may have strapped candidates into their aircraft before the flight or dealt with them in another capacity. They might say, 'Oh, yes, so-and-so was really chatty' or 'So-and-so completely ignored me' or that they 'behaved one way when a senior officer was in the room and then very differently when they left' – all of which gives us more of a 360-degree insight into the real individual and the softer areas that, as their Commanding Officer, I might not necessarily always get to see.

When we subsequently assess all the evidence from the week, our decisions are not based solely on the flight test, as is commonly believed; we don't simply go for the first-, second- or third-place pilots from the flying test. Yes, the pilots need to have exceptional flying ability and if they fail the test they will not get in, but if the best pilot displays characteristics that are not suitable for the dynamics of the team, he or she will not succeed. So if you display really good qualities and traits, but perhaps come in the middle of the pack in the flying test, then you might stand a chance of making it through. All the assessments, plus the social and sporting engagements, are there for a reason – the Red Arrows are a very tight-knit team and flying with them is hugely challenging, and there are specific characteristics that we look for in our pilots.

RED ARROWS MATERIAL

So what are the qualities of a Red Arrows pilot? As all the candidates are experienced fast-jet pilots, they will have developed the necessary skills and traits needed to fly a fifth-generation fighter, such as self-discipline, spatial sense and all-round situational awareness, intelligence and a cool head under pressure. Emotional awareness and emotional competence are key tenets for RAF leaders, and are particularly critical for team pilots in the Red Arrows. Mental resilience is also a must: the first year of training with the Reds is especially tough, and only a certain amount of tenacity is going to get you through it. Courage – a trait that all servicemen and women working on the front line must have – is also required. When you're training with the Red Arrows, the skills and techniques that you need are not gifted to you; you have to learn them yourself in close formation and at 400mph and, if you're not nailing something quickly, then you need courage to compartmentalise the issue, try again and overcome any difficulties.

In the Red Arrows, however, this blend of courage and resilience has to be balanced with a certain amount of humility. This is the primary characteristic that pilots in the team need. We are constantly assessing and evaluating our performance – a habit that has become an absolute necessity if we are to improve and achieve the high standard that we expect, year in, year out. Every sortie is videoed and dissected in minute detail in the debrief, and pilots are encouraged to point out incorrect positioning within formations or any issues with manoeuvres; and all this happens up to three times a day, with the debriefs often taking longer than the flights themselves. For this reason, pilots (Red 1 included) have to be able to take criticism and deliver it constructively, own up to any mistakes and learn from them. The whole training process more or less revolves around it.

All front-line squadrons rely on close teamwork, particularly during an operation, but there is still an opportunity to take leave away from your team or work colleagues. However, in the Red Arrows we generally take leave at the same time and are with the same ten or eleven pilots whenever we're at work. It's an environment and a process that doesn't allow for big egos. From a safety point of view, overconfidence can be a killer, simply because you may overestimate your abilities or lack consideration for fellow team members. Our process is one that encourages openness and transparency because, when flying in close formation at speeds of around 400mph, Red Arrows pilots need to have implicit trust in each other.

While there are qualities and values that the pilots share, cognitive diversity – that is, the different ways in which people think and problem-solve – is also essential for the team. We don't want a group made up of robots. Cognitive diversity is healthy for any operating unit that is making critical decisions and is looking to continually improve what it does, and various ways of thinking may stem from an individual's background or, particularly in the case of the RAF, their operational experience. In any team of pilots we might have the occasional joker, more forthright characters or those who are considered deep thinkers – all of whom can offer a different perspective, promote a questioning culture and, hopefully, help to improve our performance year-on-year. We have a long history of having New Zealanders in the Red Arrows, including original team member Ray Hanna, who was New Zealand-born; and over the last fifteen years we've had at least one Kiwi in each team, largely because the New Zealand Air Force disbanded its fast-jet force and so many of its pilots came over to the RAF. Without stereotyping a whole nation, in my experience many of the New Zealanders who join the team come with a rather relaxed, calm and measured attitude to life, which, in turn, seems to pervade the whole team, helping to counterbalance those who are perhaps a little more headstrong and making for a healthy team dynamic as a whole.

The one emotion, however, that all Red Arrows pilots share is the sheer joy they feel when they get into the team. It's an honour to be shortlisted, never mind getting all the way through. To climb into the Hawk, fly with eight other pilots at the very top of their game, then represent the RAF and the country in the hope of inspiring people across the globe, is the stuff of dreams, not to mention the most exhilarating fun.

A few days after the selection week, the shortlisted candidates are given an exact date and time for notification. As with the flying test, my memory of receiving the news is particularly vivid. On Friday 9 May 2008 I was nervously pacing around the living room of my flat in Manchester. Although my expectation of being selected on my first attempt was very low, after leaving the Red Arrows in Cyprus it had dawned on me how much I had enjoyed being around the team, and the selection week as a whole. From start to finish, it had been one of the best weeks of my career to date; and the anticipated disappointment of being told that I hadn't been successful was growing, minute by minute. At exactly 10.00 hours my Tornado Squadron Commander called me: 'Monty, good morning. I have some bad news and some good news to tell you. The bad news is I'm going to lose a squadron pilot in the near future. The good news is . . . you've been selected to join the Red Arrows as a team pilot for the 2009 display season – many congratulations.'

It took me a few seconds to process what I had just been told and to gather myself. My Squadron Commander asked if I was still on the line. I finally acknowledged him, thanked him a few times, and he said he looked forward to seeing me first thing on Monday. This news was the most impactful of all the notifications I'd received during my time in the RAF. I slowly sat on the floor in shock. From the other side of the room, my partner asked if I was okay. She thought I hadn't got in. In a highly emotional state, I uttered the words to her that I genuinely never thought I'd say: 'I'm going to be a Red Arrow.'

At exactly the same time in other areas of the UK, eight other RAF pilots were receiving similar phone calls from their Squadron Commanders, finding out whether they had been selected to join the Red Arrows. Two of them, Flt Lts David Davies (Tornado F3 pilot) and Zane Sennett (Harrier pilot), received the same happy news as me: the three of us were to report to RAF Scampton in a few weeks' time to begin training with the team. This moment would be the start of a professional and personal journey that would cement a tremendous bond between us and would set a dramatically different course in our entire careers and lives.

3

Aerial Thrills

THE ATRIUM OF THE Royal Air Force Aerobatic Team is not particularly grand – its functionality and plainness are actually rather typical of an RAF station. Along its corridors, however, are mounted boards featuring photographs of every Red Arrows team since the opening 1965 season, along with lists of the fifty-seven countries in which the team has so far displayed. For pilots and the support team, these images provide a regular reminder of the incredible history and worldwide reputation of the Red Arrows, while also prompting some to ponder on how they can build on that legacy, how they can make the team better, more global and even more inspirational for future generations.

For a pilot walking down those corridors, these boards are also a reminder that 160 display pilots before you have gone through it all previously – the huge highs of a first-rate display and the lows of a tough debrief – and that you'd better make the most of every second with the squadron as it won't last for ever. While the team is composed of some of the best the RAF can offer, each member is aware that it's the red flying suits that are famous, not the individuals wearing them, and that they're fortunate, even humbled, to have been given the opportunity to fly with the team for a while.

'You're incredibly conscious of the team's legacy,' explains recent Red 1 Martin Pert, 'and you're engulfed in that history every day. It plays very strongly on the mind – you almost feel you're being entrusted with a national asset.' The Red Arrows also form part of the well-oiled machine that makes up the Royal Air Force: they exist to showcase the skills of the RAF as a modern, capable fighting force, but also to highlight its significant role in the tumultuous history of Britain over the last century.

RAF Scampton, the long-time home of the Red Arrows, has its own remarkable story, much of which preceded the arrival of the Red Arrows at the base in 1983. Originally a First World War training aerodrome, Scampton had squadrons in action from day one of the Second World War, initially providing the base for the Hampden bombers of 49 Squadron and 83 Squadron, as well as countless other bombing raids that supported and even changed the course of the conflict. Scampton is most famously known as the base of 617 Squadron's Lancasters, which Wing Commander Guy Gibson led on the 'Dams Raid' on the night of 16–17 May 1943, breaching two of the great dams serving Germany's industrial heart.

THE PAST IS A GREAT TEACHER

Working at RAF Scampton – where Red Arrows pilots spend long days of training, technicians work round-the-clock to maintain the fleet of Hawks, and the support team plans the season meticulously – you are keenly aware of that history. The block where the 'Dambusters' aircrew gathered for their pre-sortie briefings until recently housed No. 1 Air Control Centre for many years, which provides air-control operations in the UK and worldwide, helping to protect UK airspace from hostile aircraft. The grave of Guy Gibson's dog, which was run over while

Gibson was holding a preliminary briefing on the evening before 617 Squadron launched the Dams Raid, is still situated outside his owner's former office in No. 2 hangar. And the exterior of No. 3 hangar, where the Hawks of the Red Arrows are maintained, is virtually the same as it was during the 1940s. In fact I'm often struck by how many of the buildings of Scampton look identical to those featured in the 1955 film *The Dam Busters*.

Even away from Scampton and when airborne, I'm constantly reminded of what has gone before. As a Tornado F3 pilot, I used to receive the same radio calls that Fighter Command would have used at its headquarters, Bentley Priory: 'Climb to Angels 27', meaning 27,000 feet – shortened acronyms to direct time-critical intercepts still being very much part of the RAF's lexicon today. Only recently it struck me, as we were crossing the North Sea while flying to Finland with the Red Arrows, that before we turned north to head towards Denmark and Sweden, we were steering towards Den Helder, a prominent navigation point on the Dutch coastline, one that would have been used extensively by Guy Gibson and his squadron during their early bombing raids from Scampton.

Indeed, in terms of prominent navigational features, RAF stations in Lincolnshire, such as RAF Scampton, Waddington and Coningsby, have a symbiotic relationship with the majestic Lincoln Cathedral, which, at one point in its history, was the tallest building in the world. Whether returning from a mission in a Typhoon to Coningsby, in an E-3 Sentry AWACS to Waddington or leading the Red Arrows Hawks back to Scampton from a UK or overseas task, spotting the cathedral never fails to offer a sense of being home. It is also not lost on any modern-day RAF crew how heightened and monumental that sense of relief must have been for crews recovering to the Lincolnshire airfields, returning home from their relentless and highly dangerous Second World War missions.

RAF Scampton continued its strategic importance beyond the war and into the Cold War period, this time providing a base for Avro Vulcan bombers carrying Blue Steel nuclear weapons, and there are still secure storage facilities on the airfield that once housed some of the most potent atomic weapons of the day. To facilitate the Vulcan, the runway at Scampton was extended to be some 9,000 feet, and it's a sobering thought that our Hawk jets roar up the same runway that saw nuclear-armed Vulcan bombers launch into the sky for more than thirty years. Since the Red Arrows' inception, the squadron have been parented by, and reported to, its higher headquarters within the Central Flying School. However, due to a consolidation surrounding the governance structure of the Hawk T1 aircraft, recently the team has moved into the RAF's No. 1 Group headquarters. Although this was a mostly administrative move, further change will see the team relocating to a new base, RAF Waddington, by the end of 2022. Scampton will be deeply missed, not only by the Red Arrows, but also by many aviation enthusiasts around the world. However, the team has and must evolve with the times, just as the RAF has, and wherever the Royal Air Force Aerobatic Team settles, it will still adhere to the core principles and ideals of its founder members.

Much of what we do at the Red Arrows encompasses procedures set up by the first team back in 1964, then operating out of the Central Flying School's base of RAF Little Rissington in Gloucestershire. Many of the technical aspects of what we do today – how we manoeuvre between our formation shapes and sequences, the radio calls we make to each other – date back to those founder team members and to earlier techniques of formation and aerobatic flying. In fact we have an order book entitled the *Display Directive*, which is ultimately a guide on how to be a Red Arrow. It chronicles our Standard Operating Procedures (SOPs), some elements of which have stood the test of time and date back not only to the beginning of the team, but to the early 1950s.

For a unit that is retraining pilots on a rolling basis, this kind of distilled information – what works and what doesn't – is incredibly useful and is an important element of our procedures and planning. In fact every year we hold a conference for former team leaders, during which they come to the base for the day to spend time with the team. We aim to fly with two or three leaders from different eras, so that they can pass on their knowledge and reintroduce us to things that we might have forgotten about. They may have trialled things in the past that didn't work out, and it's important we take on board some of those nuggets of wisdom so that the team can continuously improve, year-on-year.

In my first year as team leader I flew with Group Captain Ian Dick (retired), who was Red 1 between 1972 and 1974 and, at seventy-two years old, was still fit enough to pass the medical to fly a fast jet (which is impressive in itself). With Ian in the back seat, I flew an entire practice display in nine-ship formation, after which he informed me that I used the same technique to enter the formation barrel roll that legendary Squadron Leader Ray Hanna had taught him back in 1969. This form of feedback was incredibly reassuring at the time and felt like a direct link to the knowledge and methods used at the very beginning of the team's history. In Ian Dick's era the team was flying Folland Gnats and we now fly the BAE Systems Hawk T1, but the raw techniques of flying both jets are very similar. In fact the frames of some of our Hawks are more than forty years old and, like the Gnat, are a more manual type of jet when compared to the computer-driven fifth-generation fighters of today. That simplicity makes the aircraft such a joy to fly – with an enthusiasm that, for the likes of Ian Dick, has never waned – and unites team members past and present.

That love of the job and the ethos of the team have remained more or less constant throughout the decades, and the qualities that we require from our pilots today are much as they were for RAF aerobatic teams of previous decades, as outlined below by Squadron Leader Bernard Noble in 1955 for the 54 Squadron Hunter aerobatics team at Odiham: 'A display pilot should be reasonably experienced with a minimum total of at least 750 flying hours and have a flying assessment of 'above the average' on the trainer or fighter type to be flown. His formation flying must be smooth and accurate. He must be able to integrate with the team and be well disciplined and reliable. The other team members rely on him. Finally and perhaps most important, he must enjoy the job and want to do it.'*

FLYING IN FORMATION

Close-formation flying and aerobatic manoeuvres are, of course, central to what the Red Arrows do, showcasing the kind of skills and tactics that pilots have utilised since the advent of early flight. Formation flying – defined as two or more aircraft travelling and manoeuvring in a synchronised manner – was found to be particularly effective during the First World War, the first conflict in which aircraft were used on a large scale. From 1915 Zeppelins had been dropping incendiaries on England, and by 1917 biplanes were raining bombs too, with pilots increasingly engaging in aerial combat. As skies became the theatre for battle, lone fighter patrols soon gave way to groups (or flights) of aircraft operating together, evolving into larger formations made up of armed fighters escorting bombers and reconnaissance aircraft towards and over enemy territory.

Such formations, as is the case with the Red Arrows, require a flight leader, whom the other pilots in the formation (known as 'wingmen') follow, keeping their aircraft in a constant position relative to the leader's aircraft. At the point of combat, formations would inevitably break up into individual dogfights, during which aerobatic manoeuvres, designed largely to evade the enemy, gave pilots a distinct advantage. By the time of the Second World War large formations of aircraft proved less effective and cumbersome when faced with faster and more technically advanced aircraft. The Finger Four formation – consisting of two pairs of fighters, with each pair containing a leader and a wingman – soon became the

---

* We now require double the number of front-line flying hours and the assessment of risk informs much of what we do today, but the desired skills and traits are strikingly similar.

more favoured tactic of both the Luftwaffe and the Allies, as pilots could react more quickly, protect each other in the smaller formation and monitor the skies for enemy aircraft, rather than focus solely on their leader as they would in a larger, tighter formation.

With the skies becoming the theatre of battle, war increasingly demonstrated that air power had an essential and separate role in modern warfare. In recognition of this, the separate service of the Royal Air Force (absorbing the Royal Flying Corps and the Royal Naval Air Service) was created on 1 April 1918. In the closing months of the First World War the RAF carried out independent operations using a force of heavy bombers over France and Germany. Within months the RAF was the biggest air force in the world, with nearly 400 operational and training squadrons and 22,647 aircraft.

Thereafter formation flying was considered essential training for fighter pilots and they regularly practised many of the combat and aerobatic techniques developed during the First World War. A number of pilots formed unofficial teams within their units and were able to display their skills in air displays, which had been drawing large crowds for some years. Since 1912 the English aviation pioneer Claude Grahame-White had been holding weekly air displays at Hendon Aerodrome (now the RAF Museum), and in that same year 45,000 ticket-paying spectators and millions of people across London watched aviators race around London in the first Aerial Derby.

War served only to heighten the interest of the public, already captivated by these wondrous new flying contraptions, with newspapers and books delighting in the daring feats of pilots, feted as 'aces' – the A-listers of their day – who defended Britain from the horrors of aerial bombing and duelled like knights in aerial dogfights. In reality, wartime flying was harsh, aircraft were unreliable and life-expectancy was terrifyingly low, engendering a certain stoic and nonchalant humour amongst aircrews. Nevertheless, air racing and record-breaking feats further stoked the public's imagination as pilots sought to push further at the boundaries of flight, such as a non-stop crossing of the Atlantic in a Vickers Vimy bomber, performed by two former RAF aircrew in June 1919.

The end of the First World War led to major cuts to the RAF and, by March 1920, its 399 squadrons were stripped down to just twenty-nine by the then Secretary of State for War and Air, Winston Churchill. In the face of such uncertainty and searching for a means for its survival, Air Marshal Sir Hugh Trenchard, 'Father of the RAF', quickly realised the value of public displays as his best and only way of showcasing the skills and effectiveness of the RAF, and so he encouraged the remaining squadrons to form their own display teams. An idea was formed to build on the popularity of the Hendon shows by holding an official RAF 'Aerial Pageant' there, which would also raise money for RAF charities. Held on 3 July 1920, the first pageant was advertised by *The Times* as being 'in aid of the Royal Air Force Memorial Fund', featuring more than a hundred aeroplanes:

> Specially chosen RAF pilots will give an exhibition of formation flying, after which there will be a representation of one of those breathtaking melees or 'dog-fights', as they are called, in which formation is lost, and machines fight out a series of isolated combats. Then well-known air-fighters, heroes of many a combat, will re-enact, in every detail save their grim ending, a series of aerial duels . . .

The pageant featured Bristol Fighters, S.E.5s and a team of five single-seater Sopwith Snipes, piloted by instructors from the Central Flying School at RAF Wittering and led by the Canadian-born fighter ace Flt Lt Joseph Stewart Temple Fall, as they demonstrated close-formation flying and precise manoeuvres.

A second pageant followed in 1921, this time attended by King George V, Queen Mary and Queen Alexandra, which firmly established the pageant as a fixture of the London Season, comparable to attendance at Royal Ascot or Epsom races. By 1922 annual aerial shows at Hendon drew crowds of up to 195,000. Subsequent displays introduced elements employed by the Red Arrows today, such as radio commentary, which was used in 1925 and 1926 on a public-address system to inform the crowd of feats performed by a team of nine Gloster Grebes of 25 Squadron. Coloured smoke was used for the first time in 1929 to enhance aerobatic displays and became a popular feature in subsequent RAF shows – 19 (F) Squadron from Duxford fitted 'smoke-making' apparatus to its Bristol Bulldogs and, at a 1934 display, used the RAF colours of red, white and blue. At the same time the public clamoured for ever more aerial thrills, including ground attacks and bombing displays, vast fly-pasts made up of hundreds of aircraft, and even tying aircraft together with bunting.

By 1937 an RAF aerial show at Hendon drew a crowd of nearly 200,000, although this would be the final RAF display there, with the RAF instead supporting a series of displays that were held at various venues around the UK and the British Commonwealth (including Australia, New Zealand and South Africa) after 1934, known as 'Empire Air Days'. Organised with the Air League of the British Empire, these displays were designed to create greater interest in aviation and to show the RAF 'at home' at their different stations, where visitors could see how the pilots, signallers and other specialists were trained for air defence and warfare. By 30 May 1939 more than one million people were lured by the promise of dazzling displays of aerobatics, formation flying and imitation air fights at Empire Air Days held at sixty-three RAF stations and sixteen civil airfields across Britain.

While official displays were put on hold during the Second World War, millions of Britons could witness the RAF at work, as Spitfires and Hurricane fighters battled with the Luftwaffe's Messerschmitts in the skies above England during the Battle of Britain in 1940, while formations of heavy Halifax and Lancaster bombers – some of them from RAF Scampton – set off towards enemy territory, with fewer of them returning. The sights and sounds of that aircraft, with the throaty rumble of the Rolls-Royce Merlin engine, or the looming shadows of German bombers filling the sky would evoke fear, excitement and admiration for the pilots who played such a critical role in defending the nation.

ON WITH THE SHOWS

The end of the war saw the usual reduction in RAF funding, and once again the need returned for the public display of skills seen in the Royal Air Force. It was decided to absorb the Empire Air Days format into Battle of Britain At Home Days, during which the public were allowed access to RAF bases. Ninety-three RAF stations opened their doors on 15 September 1945, the first post-war Battle of Britain Day, and visitors were treated to formation fly-pasts, ground displays and solo aerobatic displays. By the late 1940s the advent of new jet aeroplanes – the de Havilland Vampire and the Gloster Meteor – further turbocharged the public's interest in air displays. By the early 1950s more than a million visitors visited RAF stations, which incorporated eight Meteor teams, five Vampire teams, two Hunter teams and a Canberra formation drill team from RAF Scampton. Meanwhile the Farnborough Air Show in Hampshire swiftly grew, from its first event in 1948, into a major international trade show where British manufacturers could showcase the best of aviation and host spectacular displays. Here the first jet airliner, the de Havilland Comet, was shown in 1949, followed by the enormous Bristol Proteus airliner in 1950. The RAF Display at Farnborough that was held on 7 and 8 July 1950 was attended by King George and Her Majesty the Queen, along with 200,000 spectators.

During the 1950s and 1960s various RAF squadron teams continued to dazzle crowds with their display teams. The mid-1950s saw two squadrons establish four-ship formation teams, with the new Hawker Hunter 43 Squadron in the North of England and Scotland, and 111 Squadron in the Midlands and the South. Led by Squadron Leader Roger Topp, the formation team of 111 Squadron expanded from four to nine, and eventually sixteen aircraft. Painted in gloss black, they were dubbed the Black Arrows (the name was coined after a French newspaper referred to them as *les flèches noir*) and were largely recognised as the RAF's premier display team. At the Farnborough Air Show in 1958 they set a still-unbroken world record looping in formation twenty-two Hunters (with the aircraft and pilots borrowed from other squadrons). Many of the manoeuvres and shapes seen today at the Red Arrows displays, such as the Diamond Nine, were flown by this very team, which included Flt Lt Lee Jones, who later went on to set up the Red Arrows.

Two years later 92 Squadron, a famed Battle of Britain Spitfire squadron, took over as the RAF's leading aerobatic team, with a team of Hunters named the Blue Diamonds. They were soon rivalled by squadrons flying the first British supersonic jet fighter – the English Electric Lightning – including 74 Squadron, the Tigers, who were followed by the Firebirds of 56 Squadron. Their days as aerobatic teams were numbered, however, as the RAF had decided that display teams should be formed out of training units and not front-line fighter squadrons, which had shrunk in number during the 1960s and needed to focus on their principal task of combat training, using increasingly sophisticated and costly fighter aircraft. Defence cuts also led to the demise of the RAF's At Home Days, although in 1961 one million people still attended the sixteen remaining RAF stations that had displays – the public appetite for such shows still going strong. The Royal Air Force Association held its first public display in June 1960 at North Weald Airfield in Essex, where the first Air Tattoo was held in 1971 and was renamed the Royal International Air Tattoo (RIAT) in 1996. Now held at RAF Fairford in Gloucestershire, this is still the largest military air display in the world and represents an important date in the Red Arrows schedule of summer displays.

The responsibility for the official RAF display teams subsequently fell to the various Flying Training Schools, which already taught formation aerobatics and had their own teams. The Central Flying School, based at Little Rissington in Gloucestershire, had formed Jet Provost aerobatic teams since 1957. By 1963 one of these, the Red Pelicans – named after its bright-red colour scheme and the pelican featured on the official CFS crest – had six Jet Provosts equipped specially for displays, including purpose-built canisters that generated smoke. By the end of that year the Red Pelicans would be nominated as the official RAF team and would continue as a display team, often alongside the Red Arrows, until 1973.

Meanwhile in 1963, at the No. 4 Flying Training School at RAF Valley on Anglesey, an instructor and former Black Arrow, Flt Lt Lee Jones, had been considering forming an aerobatic team with the RAF's new two-seater advanced jet trainer, the Folland (Hawker Siddeley) Gnat T1. The aircraft had only entered service with the RAF the year before, but it was readily welcomed by those who flew it, not least by experienced instructors like Jones, who could see the Gnat's potential for aerobatics. Future Red Arrow Dickie Duckett was a student at RAF Valley in 1963 and remembers being instructed by Jones: 'On some sorties I flew with Lee, he would arrange to meet up with the other instructors over a disused airfield near Valley and he would lead them through a few manoeuvres, so he was already thinking about setting up a team that year.'

Keen to show the manoeuvrability and speed of the Gnat, Jones took the opportunity to put on an informal four-ship display for Sir George Augustus 'Gus' Walker, Air Officer Commanding-in-Chief at Flying Training Command, when he visited RAF Valley. Sir Gus was suitably impressed by what he saw and agreed to the forming of a display team at Valley. Officially titled the RAF Gnat Aerobatic Team, it was referred to as the 'Yellowjacks' by Jones, after the Gnats' yellow colour scheme and the 'Blackjack' radio call sign used by 111 Squadron.

Consisting of seven Gnats and one reserve, the Yellowjacks went on to perform at a handful of displays in the summer of 1964. In August they put on a display at the Royal Naval Air Station at Yeovilton in Somerset, where it was thought the Gnat team had performed better than the Red Pelicans. This was relayed to the new Commander-in-Chief of Flight Training Command, Sir

Patrick Dunn, who, with some likely encouragement from Sir Richard Atherley (then military adviser to Folland, the makers of the Gnat), decided that the Yellowjacks should appear at the Farnborough Air Show the following month. Despite the short notice, the Yellowjacks again put on an accomplished display, performing a couple of manoeuvres with the Red Pelicans, followed by their own short display, before the Pelicans came back in to finish the show. The performance by the Yellowjacks was enough to convince the RAF hierarchy that the Gnat was a more attractive, exciting-looking aircraft than the Jet Provost and was well suited to precision flying and formation aerobatics.

FROM YELLOW TO RED

Running several display teams was starting to prove costly for the RAF, so in the autumn of 1964 – and as a result of the Gnat's impressive performance at Farnborough – it was decided to establish a new Gnat team under the control of the Central Flying School at RAF Little Rissington. Lee Jones would lead the team, and he and four other Yellowjack pilots were posted to the Central Flying School. Known as the Royal Air Force Aerobatic Team (RAFAT), the new permanent display team would publicise the merits and raise the prestige of UK aviation in the world market, and would stimulate recruitment to the RAF. The team was supplied with ten Gnats, five of which had been flown by the Yellowjacks and were modified with smoke canisters and VHF/UHF radios, plus a livery that featured the RAF roundels, an angled tailfin flash in the UK military aircraft colours of red, white and blue and, above this, the Central Flying School (CFS) crest. The first aircraft (XR540) was delivered on 1 February 1965 and the final aircraft (XS111) on 14 May 1965.

While the Gnat was a popular choice, the top brass at the RAF were not quite so keen on the yellow colour of the aircraft, or the 'Yellowjacks' name (the team was jokingly referred to as 'The Daffodil Patrol'). As a result, it was decided to re-spray the Gnats in postbox red – a colour associated with the Central Flying School and the CFS aerobatic team, the Red Pelicans. It therefore made sense to have 'red' as part of the team's new title, combined with 'arrows', after the swept-back, arrow-like silhouette of the Gnat, and as a reminder of the former display team, the Black Arrows of 111 Squadron. The new Red Arrows team would initially fly seven aircraft, with one reserve, and – while they had first begun training together in 1964 – the team was officially formed on 1 March 1965 at the new base of RAF Fairford in Gloucestershire, a base that was close to Little Rissington but far less busy, enabling the team to take to the sky whenever required. For its first four years the team functioned on a year-to-year basis, with its own ground crew, engineers and a Team Manager, who took overall command of the unit and was responsible for administration, arrangements for the season's programme and commentary for the air shows.

For the team's first season Lee Jones put together a team of eight pilots, all of whom were RAF instructors and some of the best formation pilots in the country. Jones himself had considerable experience of display flying, not only in the Yellowjacks and the Black Arrows Hunter team, but previously at 229 Operational Conversion Unit at Chivenor, and he had also flown Meteors, Sabres and Mosquito night fighters with 141 Squadron. The rest of the team comprised Yellowjacks members Flt Lt Gerry Ranscombe (Red 4), Flight Officer Peter Hay (Red 5) and Flt Lt Henry Prince (Red 7); Flt Lt Ray Hanna (Red 3), who was a former member of the College of Warfare's Meteor display team; deputy leader Flt Lt Bryan Nice (Red 2), who was an ex-Red Pelican; Flt Lt Bill Loverseed (Red 6), who was a former member of the Meteor display team at RAF Manby; and Flt Lt Eric Tilsley (Red 8, the reserve pilot), who was the reserve pilot for the Yellowjacks. The team also had a mascot in the form of Dusty, the golden Labrador retriever, owned by Red 7 Henry Prince, which had also been the team mascot for the Yellowjacks. Sadly, the dog was killed just before the team's first display when, after a Chipmunk had started up, he dashed out onto the runway and ran into its propeller. Dusty is buried at RAF Fairford and his grave is still tended by local schoolchildren.

Having trained over the winter of 1964 and the spring of 1965, the Red Arrows performed their first official display at RAF Little Rissington on 6 May in front of the assembled press. Three days later the team performed at the National Air Day display in Clermont-Ferrand in France, before returning to the UK to perform their first display in front of British crowds on 15 May at Biggin Hill Air Fair. There the seven-ship team thrilled a crowd of 40,000 people with a fifteen-minute display featuring tight formations and manoeuvres, including the Vixen Break (see Glossary) and low-level passes, with the synchro pair racing towards each other at a combined speed of anywhere between 700 and 800mph.

The Red Arrows went on to perform at a total of sixty-five shows that year, both in the UK and in mainland Europe (including West Germany, France, Italy, Belgium and the Netherlands), and also formed part of the RAF presentation at the Paris Air Show at Le Bourget alongside the Lightning F3s of No. 111 Squadron. During that season the Red Arrows earned acclaim from the press and public alike, and in 1965 they were awarded the Royal Aero Club's coveted Britannia Trophy for 'their meritorious service as the World's greatest aerobatic team'. It was a very promising start.

4

Éclat

WITH A SUCCESSFUL first season under their belts, the Red Arrows began training for the 1966 season, this time under the helm of newly promoted Squadron Leader Ray Hanna, who would lead the team for the next four years. Subsequent thrilling displays would see the Red Arrows further establish themselves as the world's premier jet formation team, laying down many of the manoeuvres and shapes that are common to the team today, including its signature formation, the Diamond Nine.

From the outset the team sought to keep its displays visually exciting and 'tight', with loops, sharp turns and manoeuvres moving seamlessly into the next ones – a strategy that runs at the core of Red Arrows displays to this day. Compared to the team's current steed, the Hawk, the Gnat was able to loop at a relatively low height, meaning that the Reds had more opportunities to use vertical manoeuvres to twist and keep the show centred. Teams then and now, however, aimed to keep as much of the formation and action in front of the crowd as possible, avoiding any lengthy intervals of empty sky. 'If the crowds have time to lick their ice creams,' reminded the team's charismatic leader Ray Hanna, 'we aren't doing our job properly!'

For public displays the Red Arrows flew as a seven-ship formation – the seventeen-minute show featuring an arrival loop and, back then, around twenty formation changes, including the Swan, Cutlass, Arrow and Vixen, with the team splitting into five aircraft and the synchro pair (Reds 6 and 7) performing the high-speed opposition pass, the Roulette. The year of 1966 also saw the first (although rather fleeting) appearance of the Diamond Nine formation when the team decided to bring in their two reserve pilots for a handful of special displays, with the first nine-ship display being flown on 8 July 1966 for HRH the Duke of Edinburgh at Little Rissington.

In 1968, for the Red Arrows' fourth season, it was decided to increase the team to nine aircraft on a permanent basis. The Reds could now form many different formation shapes, such as the Diamond and the Arrow, alongside more varied and complex sequences, with the team splitting off into two sections so that one could always be performing in front of the crowd. Reserve pilots were no longer included in the team, as the practicality of training pilots so that they could fly in any position – each of which has its own specific challenges – meant that they were highly skilled, but still only in reserve. Far better, and much safer, to train all nine pilots to fly in specified positions, while accepting that there might be the occasional display where there was an aircraft down, in the event that a pilot was unfit or unable to fly. The only exception to this, of course, is Red 1, who leads and orchestrates the whole display, with all the other pilots formatting on the leader's aircraft and trained on his or her every move and call – without the leader, there is no display.

By 1968 the Gnat featured, in addition to its red, white and blue tailfin, a white lightning flash on the side of its fuselage. The Folland Gnat was a fast and agile jet, described by pilots (many of whom had trained and instructed on it) as an 'absolute joy to fly'; it was a jet you didn't climb into, many of them quipped, so much as 'put on'. With a wingspan of only twenty-four feet, the Gnat was much smaller than the Hawk. 'When you sat in it,' remembers former Red 1 Brian Hoskins, 'it was like a giant go-kart, because you were right up front and it was as though you were strapping wings onto you, and the view out of the aircraft and its handling were just wonderful.' Dickie Duckett, who joined the team for the 1968 season, also has fond memories of the jet: 'Walking out to it, you could look straight into the cockpit, you didn't have to climb any ladders. It was really manoeuvrable and its forty-degree swept wings suited the Diamond-formation shape really well. Its swept-wing shape also helped it turn more tightly than a Hawk so it could fly really tight displays.'

However, the team's modified Gnats, in which two of its integral tanks were adapted to store diesel to create smoke for displays, consequently had a much lower fuel capacity and therefore a more limited range. When transiting longer distances, 'slipper' fuel tanks could be fitted under the wings, but for aerobatic displays the team often worked to very tight fuel margins, with enough for fifteen-minute displays, but with restricted flights to and from sites. 'If we were doing a display, say, at a seaside resort,' Dickie Duckett remembers, 'we'd have a range of about thirty-five miles, meaning we had enough fuel to take off, fly there, do the display and fly back. I remember on one or two occasions, when a display was more like forty-five miles away, we'd tow the aircraft out to the end of the runway so we didn't waste too much fuel taxiing out.' Thankfully the Gnat's low-level fuel warning light, on which the team was heavily reliant, was very accurate.

CALM UNDER PRESSURE

The 1968 season saw the arrival of two future team leaders, Flt Lt Ian Dick as well as Flt Lt Dickie Duckett. Dick was an advanced jet flying instructor at RAF Valley, just six months into the role when he received the call that the Red Arrows wanted him for the team, and luckily his boss at RAF Valley let him go. As an instructor on the Gnat, he flew out of the Red Arrows' base of Kemble and had also previously trained with Flt Lt Henry Prince, who was in the first Red Arrows team of 1965 and was leader of the synchro pair by 1968.

Duckett had seen the Red Arrows perform at his base in Germany, where he was flying the high-performance interceptor fighter, the Lightning. Already familiar with one or two people in the Red Arrows (including Lee Jones, who had been his instructor at RAF Valley), Dickie let it be known that he was keen to get on the team, eventually joining in February 1968. At that time the usual procedure was for new pilots to start out on the side they were going to fly in displays (even numbers on the right of the formation, odd numbers on the left). While new pilots would start close to the leader, back then they were generally placed on the outside of the formation, where it was thought they would 'cause less trouble' if they lost their position. The more experienced pilots were positioned on the inside – the idea being that they would give a more stable base to the formation. In later years this philosophy would change, and today pilots begin at the front of the formation close to the leader, flying as Red 2, 3, 4 or 5, and move outwards, and further back, in their second and third years.

By 1968 Ray Hanna was into his third year as Red 1, a leader whose flying skills, judgement and natural instinct for aerobatics earned him the total confidence of the team around him. Ian Dick certainly felt he was learning from the best – 'He took us under his wing and just told us to get on with it' – and his regard for his former boss remains undiminished to this day. 'Ray Hanna is my one and only hero. He was probably one of the (if not *the*) most outstanding RAF pilots, if not worldwide pilots, following the Second World War. He was exceptional in every way, both in the air and on the ground.'

A man of few words, Hanna led with quiet authority and by example if the need arose, as Dickie Duckett remembers: 'I would never question anything he decided to do. I had an incident early on in my first year – the whole team went to land on an airfield on Thorney Island. We were misled about the wind and state of the runway – I was the first to land and went off the end of the runway into the barrier, as did the next chap, and then everyone did. My aircraft was a bit damaged and the ground crew fixed it over the lunchtime as we had a show that afternoon, and they had to take one or two bits off it. Ray asked me: "Are you happy to fly it?" and I answered that, no, I wasn't really, to which he replied: "Okay, I'm going to go and fly it and if I'm happy, are you happy?" I told him yes, that would be fine, and I ended up flying the plane – the bits that were missing weren't vital, obviously.'

In leading the formation, Hanna gave radio calls that were calm and confident, but he tended not to mince his words, as Ian Dick experienced when flying from RAF Valley to a display at Black-pool as Red 7. His aircraft had suffered hydraulic failure, meaning that he had lost powered controls, and the flap, undercarriage and braking operations were also limited. When Dick radioed through to Hanna to inform him of his predicament, Red 1 immediately replied in his distinctive Kiwi accent: 'Red 7, no need to shout about it. Just get on with it.' Ian did as he was told and – as his boss had no doubt predicted – managed to land safely.

At the end of 1968 Squadron Leader Tim Nelson took over as team leader, having flown with other Hunter and Lightning aerobatic teams, and as Red 7 with the Red Arrows in 1966. Bad weather would hamper training for the 1969 season, and on 26 March the Red Arrows suffered a fatality when new team member and former Lightning front-line pilot Flt Lt Jeremy Bowler struck trees while flying at a low level during a practice display at RAF Kemble. The team nonetheless completed training successfully and performed some twenty-seven public displays, albeit with a few difficult moments. However, with all of the exceptional challenges felt during the pre-season work-up, the team was struggling in performance and paused its display programme after a show in Germany on 15 June at RAF Wildenrath. Nelson, in consultation with others and the chain of command, took the decision to stand down as Red 1. Ray Hanna was recalled as an interim leader. Over a single week and seventeen sorties, the team trained intensively under Hanna's leadership and then continued with a busy schedule of displays at home and in Europe.

After leading for the rest of 1969, Hanna left the team and Dennis Hazell took over as Red 1 for the 1970 season. He had previously been an instructor and was an experienced fighter pilot, with some formation aerobatic experience although, unusually for a Red 1, he had never flown with the team before. Training for that season saw the loss of two Gnats in an incident, although the two pilots involved, Jack Rust (Red 8) and Dickie Duckett (Red 4), were unharmed. The incident, as Dickie remembers it, occurred during a practice with five other aircraft. Having just changed positions with another Gnat, he was told on the radio, 'Dickie, you're on fire!' – at which point he broke away from the formation, even though he wasn't aware of any issue. The fire and smoke that the team had seen, and which Hazell had assumed was coming from Dickie's jet, was in fact billowing out of Rust's Gnat and, as his radio had stopped working, he couldn't be heard. On seeing Rust's aircraft spiral down, the leader radioed, 'Dickie, get out!' and Red 4 duly followed orders and ejected. Still unaware of the mix-up, Duckett landed on the ground just off the airfield and was taken to the crew room – 'and then Jack Rust came in, looking a bit stressed,' explained Dickie, 'and it suddenly dawned on me what had happened.' Luckily, both escaped injury, other than 'a bit of a stiff neck for three weeks'.

The 1970 season saw a number of new manoeuvres and formation shapes, including the triangular formation known as Apollo, in honour of the historic Moon landing the previous year – which the 1969 team had managed to watch after a display in Strasbourg, standing on a pavement in front of a television shop. In 1970 the team also performed the popular Twizzle, which saw the aircraft flying so that they were lined up on each other's left wing and in close succession, rolling to the right through 270 degrees and under those following. The new nine-ship formation had also enabled seven of the formation to split off and carry out some showier manoeuvres, such as the Bomb Burst, with the synchro pair largely filling the gap in the sky until the formation got back together again. Ian Dick, as Red 6, and Dickie Duckett, as Red 7, expanded on what the synchro pair usually did in terms of manoeuvring, introducing some of the more low-level dynamic passes that we see today in Red Arrows displays. They also invented a new synchro manoeuvre known as the Boomerang, which saw the synchro pair run in over the crowd, split, pull up into a loop, roll over and cross over in front of the crowd.

Today the synchro pair flies at a low altitude, but no lower than 100 feet. Fifty years ago the pair flew at lower heights, sometimes as low as fifteen to twenty feet, partly because fighter pilots were used to flying at a very low level on operations so that they were under enemy radar systems. Synchro pairs usually flew in at a height they felt was comfortable and safest for them and, as the Gnat altimeter didn't provide an accurate reading at a low level, that was all done by eye. 'People were getting a bit uncomfortable with those low levels,' explains Dickie, 'and they were quite keen to fix our height. We agreed that we would not fly below the height of the control tower at Kemble, which we were told was thirty-five feet high. Some months later it was found to be sixty-five feet, but I don't think anyone really noticed the difference.'

Dennis Hazell's tenure as leader was to end prematurely during winter training for the 1971 season when he suffered an unfortunate accident. After experiencing engine failure while practising a Twinkle Roll, he was unable to perform an emergency 'dead-stick landing' as two Gnats were blocking the runway. Forced to eject, he broke both his legs badly in the process. As a result, Bill Loverseed, who was the current Red 9 and a former Red 6 in 1965, stepped in as Red 1. The year of 1971 also saw the worst accident in the history of the Red Arrows when, on Monday 20 January, the wings of the synchro pair struck each other during a close passing manoeuvre, killing Flt Lt Euan Perreaux and Flt Lt John Haddock, along with two new pilots in the back seats, Flt Lt John Lewis and

Flt Lt Colin Armstrong. Following an investigation into the engine failure in Hazell's aircraft two months earlier, it was decreed that engine handling would be monitored and recorded by a second pilot, who would sit in the back seat of the synchro-pair aircraft during some practice sorties. That is why, unusually, there had been two pilots in each of the synchro-pair aircraft.

The accident occurred during the final cross of the Roulette manoeuvre (a low-level 360-degree turning manoeuvre with the aircraft flying in opposite directions). Red 6, Euan Perreaux, during the inside of the turn and belly up to Red 7, John Haddock, lost sight of him and, thinking that he had passed him, started to roll out of the turn just as Red 7 was pulling towards him in an effort to make a closer cross, and they collided. The accident was devastating for the families involved and was covered extensively in the national press. The Red Arrows received a great outpouring of sympathy, with the public being even more appreciative of the considerable skills demanded by the team and strongly urging them to continue, although questions were raised about the viability of the team. The Red Arrows pressed on with a seven-ship formation, performing seventy-one displays in the UK and continental Europe without further incident that season.

Ian Dick, now Squadron Leader, took over as Red 1 for the 1972 season, mindful of the advice he'd been given by Ray Hanna that a Red Arrows display must, as the plain-speaking team boss put it, 'thrill the ignorant, impress the knowledgeable, but frighten no one'. Not long after he had secured the job as leader, Ian was informed by his commander that he was to take the team to Canada and the United States. But taking a fleet of single-engine Gnats across the icy waters of the Atlantic was an ambitious task. 'Operation Longbow', as it was called, ultimately proved extremely successful, with the team performing in front of large crowds, including at the world's largest air show in Washington, with a watching audience numbering upwards of 1.2 million. The subsequent year saw the team perform 100 displays across the UK and mainland Europe, and Ian Dick arranged for the team to fly in formation with the iconic supersonic airliner Concorde, producing some stunning photography in the process.

At the end of the 1973 season Squadron Leader Peter Squire, a former QFI and Hunter display pilot, joined the Red Arrows as team leader. Not having flown with the Reds before, he had a difficult act to follow after Ian Dick and was further hampered by poor winter weather and a fuel crisis that limited flying training. By the early spring in 1974 it became clear that the team would not be ready to display under Peter's leadership, so it was decided that Ian Dick should be brought back to lead the Red Arrows for the season, which began in late June and totalled fifty-nine displays. Sir Peter Squire (deceased) went on to fly Harriers with No. 3 Squadron in West Germany before being promoted to Commanding Officer of No. 1 (F) Squadron, taking over from (OC) Dickie Duckett in 1981. The following year he led the squadron on Harrier ground-attack operations in the Falklands War, where he was in the thick of the action and was later awarded the Distinguished Flying Cross for valour and leadership. He was subsequently knighted and became Air Chief Marshal Sir Peter Squire, UK Chief of the Air Staff.

Squadron Leader Dickie Duckett took over as leader of the Red Arrows for the 1975 season, a time that saw a fair bit of changeover, with four new pilots also joining the team and John Blackwell replacing existing team pilot Mike Phillips during the season. One new team member, Squadron Leader Brian Hoskins, took on the position of Red 7 – part of the synchro pair, a position normally reserved for pilots in their second or third year of the team, although the former fighter pilot had clearly shown his mettle in leading the Macaws aerobatic team between 1971 and 1972. Brian was delighted to be led by Flt Lt Des Sheen (former Red 8), with team leader and former Red 7 Duckett also passing on invaluable advice. On 18 April 1975 the Red Arrows flew behind an Avro Lancaster at RAF Kemble, the team forming the shape of a Lancaster in tribute to the crews who flew in this aircraft during the Battle of Britain – a poignant moment for the team leader, whose uncle had been killed in the Second World War bomber.

The hot summer of 1976 saw the team fly 90 per cent of their shows as full displays, meaning that the formation could fly full loops, as opposed to the flat and rolling displays when, as is often the case in the chilly climes of the UK, the clouds are lower than 5,500 feet. The aim of the team has always been to put on the best possible display, substituting loops with rolls, sharp turns or fly-pasts, so that even on cloudy days it can give an impressive performance and even switch between different displays and sequences as it flies.

But while the hot weather of 1976 allowed for more impressive displays, it inadvertently led to leader Dickie Duckett losing his aircraft, number XS111, a Gnat he'd flown for four years and more than 1,100 sorties. Team members of the Red Arrows (and that includes engineers) often feel a real sense of connection with the aircraft to which they entrust their lives. Flying them two or three times a day, with the pilot and Circus-engineer names painted on the side, they become attached to the particular 'character' of their jet, its foibles and its subtle differences in handling. After a solo training sortie in June 1976, Dickie landed and discovered that only one of his brakes worked, an issue for the Gnat, which relies on differential braking rather than nose-wheel steering to go straight on the runway. 'Without any right brake, I could only go left across the airfield, and the ground was so hot and very dry that I wasn't slowing down. With the boundary wall looming, I had to pull the undercarriage up, which stopped me very suddenly.' Dickie was fine, but his dear friend XS111 was damaged beyond repair.

By 1976 the selection process for the team no longer required all its pilots to be former flying instructors, although applicants were still expected to be assessed as 'above average' by their commanders and to have at least one operational tour under their belt. In 1977, with Squadron Leader Frank Hoare leading the team, the Red Arrows played a key role in events commemorating the Queen's Silver Jubilee. The subsequent year saw tragedy when, during a pre-season practice, Flt Lt Stephen Noble and former team leader Wing Commander Dennis Hazell, in his back seat, were killed as they practised Rollbacks at a low altitude – a manoeuvre that sees pairs of aircraft pull up, perform a barrel roll and re-join the formation – his aircraft becoming inverted as it impacted the ground.

FROM GNATS TO HAWKS

The 1979 season would see the return of former Red 6, Squadron Leader Brian Hoskins, as team leader, following a tour flying Buccaneers as a Flight Commander on 208 Squadron. It was a straightforward, busy last year for the Gnat, but was not without some problems. By the late 1970s the Gnats, although they had served the Red Arrows well, were beginning to show their age. The team was the last unit to be flying the aircraft, and the engineers were finding it difficult to source spare parts and effectively maintain them. The decision was therefore made to replace the Gnat with the RAF's latest advanced jet and tactical-weapons trainer, the Hawker Siddeley (later British Aerospace and now BAE Systems) Hawk T1, which was already in use at air-force training schools across the country. The two-seater Hawk had proved itself a strong and rugged trainer, being bigger than the Gnat with a wingspan of 30 feet, but still highly manoeuvrable. Its Rolls-Royce/Turbomeca Adour 151-01 engine was reliable and fuel-efficient, with a far greater range than the Gnat, offering more fly-pasts and events in the same day. Significantly, the greater range of the Hawk meant that RAFAT could now embark on achieving a more global footprint, creating further opportunity to support strategic campaigns all over the world.

Hoskins first needed to prepare the team for its final season flying the Gnat, although one pilot, Martin Stoner, had to have an operation and missed the first three months of training. As he was going into his third year with the team, it was decided to continue training without him. By early January Martin had returned to flying and the team was cleared to display on 2 April. The display season started well and included both familiar and new venues. Unfortunately, after forty-two displays another pilot was diagnosed with leukaemia and was forced to leave the team. The initial decision was that the Red Arrows should reduce to seven for the rest of the year, meaning that no retraining would be required and no displays cancelled. However, Hoskins managed to convince his bosses that he could retrain the team to fly as an eight-ship formation without cancelling displays. Red 8 moved to number 4 and Red 7 flew in the number-8 position for the first half of the display. After seven practices the team was ready to display and saw a particularly good performance by the new Red 4, Neal Wharton, who had to perform some manoeuvres that he hadn't flown before.

In preparation for the transition to the new aircraft, during 1979 Hoskins was converted to the Hawk by the most senior Hawk pilot at Kemble, Wing Commander Ernie Jones. The British Aerospace factory at Bitteswell in Leicestershire, which did the winter servicing of the Gnat, was tasked with preparing the fleet of Hawks for the team. The only modifications made to Red Arrows Hawks were the fitting of pods, piping and switches that, via the jet exhaust, generated the red, white and blue smoke; and a modification to the fuel-control unit. This gave the pilots a slightly faster response to throttle changes, enabling more precision in manoeuvres. The factory also painted the aircraft: white flashes were added on the underside of the red Hawks, together with a new badge featuring the trademark Diamond Nine formation, and the team's motto, '*Éclat*' – from the French for 'excellence'. By the middle of August the first Hawk was ready for delivery to Kemble. The team leader took the Gnat team to Bitteswell. After a formal handover of the Hawk, Hoskins led the team back to Kemble, with Wing Commander Ernie Jones in the back seat of his plane.

On 15 September 1979 the Gnat, having flown 1,292 displays and visited eighteen countries with the Red Arrows, would perform its last-ever show at RAF Valley, which was fitting as this was where the team's founder initially conceived of the Red Arrows team. By the middle of October twelve Hawk aircraft had been delivered to the team, ready for them to practise their routines for the next season. Two new pilots, Byron Walters and Tim Watts, had joined the Reds. Byron was a Hawk instructor at Valley and he trained Tim, while the rest of the team went to RAF Valley to convert to the Hawk.

Brian did some early formation flying with the new pilots and began to fly the planned display for the team on his own. In his own words: 'My starting point was to fly in exactly the same way as I'd flown the Gnat the year before. I wanted to get as close to the display that had been developed by Ray Hanna and to fly the same tight pattern. A lot of people did not believe that nine Hawks in formation would be able to stay as close to others in a tight display. I found, from flying it on my own, the pattern was much the same as the Gnat. I used a little more height in loops, went a little bit wider coming back in, but by the time all the pilots returned to Kemble, I was able to fly much the same display as we flew before.'

An immediate benefit of the Hawk was than its engine was a lot less thirsty and carried far more fuel than the Gnat. The Gnat, with full fuel tanks, could fly one-and-a-half display practices, whereas the team had to limit how much fuel they put in the Hawk to fly two practices. While Brian put together a display that delivered just as much as the Gnat had done, he made a few changes, splitting the formation into a front five and back four, rather than a seven-ship and synchro-pair split, bringing back the Twinkle Roll when the synchro pair had split off and adding a few new elements to lengthen the show to around twenty-four minutes.

As the Hawk was a larger aircraft, when the formation rolled or changed shape, those on the outside had to fly a greater distance or height than those on the inside. As a result, pilots on the outside of the formation needed to anticipate the changes; to help them, Hoskins gave the team earlier warning of the changes over the radio so that they could anticipate the moves. Brian also purposely kept together the 1979 synchro pair, Flt Lt Richard Thomas and Squadron Leader Steven Johnson, because he knew they would have a lot of work to do on their patterns and manoeuvres. When pulling G, both in the Gnat and the Hawk, a useful indication that the aircraft is at maximum lift is felt by a gentle vibration in the airframe, known as 'buffet'. Light buffet tells the pilot that the aircraft is offering the maximum turning performance possible. When pulling G in the Gnat, the pilot can feel a wide range of buffet and, by matching the buffet in the Gnat, at a set speed, the synchro pair can match their rates of turn, which is especially important when they cannot see each other. Because the Hawk has a much straighter wing and is not swept, it does not have the Gnat's range of buffet, and so the synchro pair needed to find other ways to match their turns, especially in their patterns, if they were to meet at their planned crossing points.

The arrival of twelve brand-new Hawks made life much easier for the whole team – in the previous seasons, between September and January when the aircraft went off for winter servicing, the team might have had six or seven aircraft available for training. In contrast, the team now had twelve aircraft at their disposal during the winter of 1979, meaning that they could build up the formation to a full nine much earlier than they had done on the Gnat. The arrival of the Hawk also coincided with the start of winter training at RAF Akrotiri in Cyprus, initially for two weeks. Over many years the Gnat team leaders, because of the shortage of aircraft, winter servicing and poor weather, could really have done with such a detachment. Nonetheless, the 1980 team enjoyed the glorious weather and flew displays at RAF Akrotiri and the Sovereign Base Areas.

The first few displays of 1980 lived up to the '*Éclat*' emblazoned on their new aircraft, although the season was not without incident. During a display at Brighton on 17 May, Red 7, Squadron Leader Steven Johnson, clipped the mast of a yacht, which had ignored a ban in the area and had motored underneath where the synchro pair was performing – and a yacht without sails is difficult to see. The mast took a couple of feet off his wing and the aileron, meaning that Johnson had no control of his aircraft and was forced to eject over the sea. Despite leaving the aircraft as it rolled sideways and at a low level, the Martin-Baker Mk10 seat, a design originally developed for the Harrier, rocketed him up and away from the aircraft. After ejecting, Steve was picked up by a helicopter, while Brian took the rest of the team to Biggin Hill as they had another display to fly. 'We were sitting on the grass, doing the brief for the next display,' remembers Brian, 'when we saw Steve being dropped off from a helicopter, still soaking wet. I shouted over to him: "Steve, you're just in time, we've got a spare aircraft" – at which point he looked a bit shocked, until it dawned on him that I was joking. We were all relieved to see him safe and well, so took off and did a good eight-ship display. Steve had a few days off and by the end of the week we were back as a nine-ship formation.' In the Hawk's first year the team flew 119 displays.

In 1981 the Red Arrows left training in Cyprus to conduct a tour in the Middle East, supported by British Aerospace and other manufacturers, which saw healthy international orders for the Hawk. The summer season included fly-pasts marking the opening of the Humber Bridge and the wedding of Charles, Prince of Wales to Lady Diana Spencer. Squadron Leader John Blackwell took over as leader between 1982 and 1984 and would oversee the team's move from RAF Kemble to RAF Scampton in Lincolnshire in 1983, which was home to RAFAT's parent unit, the Central Flying School. Apart from a short period at RAF College Cranwell, the Lincolnshire station has been the Red Arrows' home ever since.*

A NEW WORLD

The Hawk's greater range enabled the team to travel more widely, as it continued to do under the leadership of John Blackwell with a second tour to North America in 1983, after it had flown in formation with three of the RAF's new fighter aircraft, Tornado GR1s. The North America tour saw the Red Arrows fly with the US Air

* The Red Arrows were expected to relocate to RAF Waddington by the end of 2022.

Force Thunderbirds and with the US Navy aerobatic team, the Blue Angels – the Reds performing to huge audiences and ably demonstrating the skills of British pilots and aircraft. Squadron Leader Richard Thomas returned to the team as Red 1 in 1985, and the following year he led it on an extensive tour of the Gulf and Asia-Pacific region, stopping in nearly twenty countries, including Jordan, India, Indonesia and Singapore. Along the way the Reds dealt with forty-five-degree heat and tropical storms, with the robust Hawk – with the help of a dedicated team of engineers – coping with all that was thrown at it.

During the second half of its displays the team now split into a four- and five-ship formation, as introduced by former team leader Brian Hoskins. The front five aircraft, made up of Reds 1 to 5, were (and still are) referred to as 'Enid', after Enid Blyton, author of *The Famous Five* books – with Red 5, a second-year pilot who mentors the new pilots, often being known as 'Uncle Enid'. The back section is made up of Reds 6 and 7 – known as the synchro pair – and Reds 8 and 9 who, collectively, are usually the most experienced pilots in the team.

Training for the 1988 season was badly hampered when, in November of the previous year, the fin of Red 2, Flt Lt 'Spike' Newbury, struck Squadron Leader Tim Miller's Hawk, causing both pilots to lose control and eject; one of the Hawks crashed into a house in the nearby village of Welton, but fortunately no one was hurt. Spike broke his femur badly and Tim damaged his back, but, despite being told by doctors that he might never fly fast jets again, he managed to return to the team fairly rapidly. Tragically, three months later in January 1988, Red 8, Flt Lt Neil MacLachlan was killed while the team was practising Rollbacks. In June that same year, Red 4, Squadron Leader Pete Collins, was forced to eject when his aircraft struck the ground during a low-level pass at RAF Scampton.

With the fall of the Berlin Wall in 1989 and the end of the Cold War, the team was for the first time invited to visit Russia, Ukraine and Hungary in 1990, marking Squadron Leader Tim Miller's final year as Red 1. In 1991 the Russian Knights display team was a guest at Scampton, during which it performed a fly-past with its British hosts over the Balmoral estate, the Scottish home of the Royal Family. Squadron Leader Adrian Thurley's tenure as leader from 1991 to 1993 saw busy campaigns in the UK and abroad, including a new tour of North America in 1993. The team also flew with the last remaining airworthy Avro Vulcan bomber, its large iconic delta-wing shape tucked in behind the nine-ship formation. And in 1991 the team was visited by the actor David Jason, who flew in the back seat of Red 1's aircraft. During his tenure as boss, Thurley also changed the rotation of pilots, putting new pilots in as Red 2, 3, 4 or 5 rather than on the outside of the formation.

Former Red 6, Squadron Leader John Rands, returned to lead the team for the 1994 season, during a period when the Cold War was at an end and the Red Arrows were once again facing scrutiny by the RAF/MOD, which were under pressure to cut costs. In every strategic defence review the validity of the RAF and HM Forces capabilities is scrutinised, primarily to make sure that the MOD has the best balance of people and equipment to defend or promote national interests. Although they have a proven strategic 'soft-power' capability and are the most visible UK military asset, the Red Arrows have often come close to being disbanded in order to save money.

Rands and his seniors were asked to consider a range of questions: could the team fly Tucanos instead of Hawks? Or fly a seven-ship formation instead of nine? The final assessment was to leave everything well alone. Emphatically demonstrating their contribution to international defence engagement, the Reds toured to the Middle East in 1995 and then, for the first time in the team's history, to South Africa and Zimbabwe followed by the Middle East and Malaysia, where the team returned in early 1996 before heading to Australia. On 26 January, coinciding with Australia Day, they performed to huge crowds at Sydney Harbour and this remains one of the biggest single audiences to have seen the Red Arrows carry out a live display.

Squadron Leader Simon Meade took over as Red 1 for the 1997 season, a year in which Virgin Trains named one its locomotives *The Red Arrows* and the team toured to the Middle East, Pakistan, India and Malaysia. Two years later the team carried out a fly-past over Edinburgh to mark the opening of the Scottish Parliament, and a fly-past over the Forth Bridge in formation with a regular 'Red 11', Concorde. The 1999 tour to India, Malaysia and Egypt ended a successful three-year tour for Meade, who in July was promoted to Wing Commander, becoming the first Red 1 to hold the rank, and he was subsequently awarded an OBE. The year 2000 saw the team carry out, under the helm of Squadron Leader Andy Offer, special fly-pasts for the 100th birthday of their honorary commandant, Queen Elizabeth, the Queen Mother, and for the last football match at the old Wembley Stadium: England versus Germany on 7 October. At the end of the year RAFAT moved back to their old RAF base of Scampton, the previous decision made by the MOD now being reversed. Squadron Leader Spike Jepson took over from now-Wing Commander Andy Offer for the 2002 season, during which the team did a special fly-past for the Queen Mother's funeral and took part in a fly-past over Buckingham Palace with Concorde, joining the formation to mark Her Majesty the Queen's Golden Jubilee. Pictures from this spectacle were transmitted around the world.

The fortieth display season of 2004 was marked by a number of special events as well as a tragedy, when Red 2 Matt Jarvis ('Jarvo') was diagnosed with cancer and sadly lost his life the following year. The 2005 season saw the arrival of a new leader, Squadron Leader Dicky Patounas, and a special performance at Biggin Hill airport to mark the fortieth anniversary of the team's first UK display, as well as a display on the Solent near Portsmouth for the Royal Navy's International Fleet Review, as part of the 200th anniversary of the Battle of Trafalgar. The team flew their 4,000th display at RAF Leuchars on 9 September 2006, following an extensive tour in mainland Europe, Asia and the Middle East, and a fly-past marking the Queen's eightieth birthday over Buckingham

Palace with the RAF's last Canberra PR9, after fifty-five years in service.

THE MODERN REDS

Under its new leader, Wing Commander Jas Hawker, the team unveiled a revised livery in October 2006 with the words 'Royal Air Force' added to both sides of the fuselage. The next month an extensive tour to the Middle East, India and Malaysia saw the team fly in unusual formations, including a fly-past in Battle formation with a General Dynamics F-111 of the Royal Australian Air Force tucked in and following, with its giant fire-plume trailing behind. On 1 April 2008 four Typhoons flew over Tower Bridge in London with the Red Arrows to mark the ninetieth anniversary of the RAF, and in June of that year the team travelled to Canada and the USA, meeting up with their old friends the Thunderbirds, the Blue Angels and the Royal Canadian Air Force Snowbirds.

The season marked the first year of Mike Ling, Red 3, with the team and he looked forward to visiting the United States: 'The first time I set foot in the US was in a Red Arrows jet and it was brilliant flying down the east coast. We started at Langley Air Force Base in Virginia, home of the F-22 Raptor, at a huge air show where we were very popular, as it was the first time the Reds had displayed there. The icing on the cake was flying around New York Harbour. We had special permission to fly past the Statue of Liberty and down the Hudson River, but did it at a very low level, so we were able to go around the Statue of Liberty (not over) and below the Empire State Building.'

Later that year, once the team was back from North America, I joined it, flying in the back seat for the latter part of the display season, before beginning training for the subsequent season in the Red 3 position. The display season of 2009 saw many standout moments, including a display over the Rock of Gibraltar as part of its Battle of Britain commemorations. This was particularly memorable for me as a bird hit my aircraft just before I broke formation to land, with the bird colliding into the metal probe – the pitot-static tube – at the front of the Hawk, almost taking out the front underside of the plane. I managed to land okay, but the jet was stuck in Gibraltar for several weeks.

Former Red 6, Ben Murphy, was made Team Leader for the 2010 season and the year also saw the arrival of the team's first female pilot, Kirsty Murphy, as Red 3. I was now part of the synchro pair as Red 7, with Flt Lt Mike Ling leading as Red 6. Springhawk training went ahead as usual in March 2010, the plan being to train for two weeks at the Kastelli Air Base in Crete before heading on to Cyprus. Whilst practising a manoeuvre in Crete, Mike Ling and I were involved in a collision, forcing Mike to eject while I managed to land my Hawk. 'Lingy' suffered multiple injuries and was unfortunately out for the season, while I was able to re-join the team a few weeks later, with Flt Lt Pablo O'Grady replacing Mike as Red 6. The team went on to perform seventy-two displays in the UK and mainland Europe.

In 2011 I flew as synchro leader and the team enjoyed a good work-up in training, then progressed to a display season, with shows in Greece, France, Slovenia and Turkey as well as displays across the UK. During a three-day display at Bournemouth the team suffered tragedy when on 20 August Red 4, Flt Lt Jon Egging, was killed when his Hawk crashed as he broke from formation, ready to land at Bournemouth airport after a beachfront display. And the team would again be grounded as it prepared for the 2012 season when, at RAF Scampton on 8 November, Red 3, Flt Lt Sean Cunningham, was ejected from his cockpit while conducting pre-flight checks on the runway. His parachute failed to deploy and he died from multiple injuries.

Squadron Leader Jim Turner had been in the role of team leader for just three weeks when Cunningham was killed, and he now had the task of leading the team during one of the most difficult periods of its history. He had to deal not only with the emotional impact of the team losing two of its pilots, but also with scrutiny from the senior chain of command, ongoing service inquiries, delays and interruptions in training and fewer serviceable aircraft – often only four or five – on which the pilots could train. In early 2012 Flt Lt Kirsty Murphy left the team due to personal circumstances, and the decision was made that RAFAT would fly as a seven-ship formation, other than for special fly-pasts when Red 8, Dave Davies, and former Red 6, Mike Ling – now recovered from his injuries and back with the team as Red 10 – would make up a nine-ship formation.

The summer saw plenty of high-profile events, including a series of fly-pasts to mark the Queen's Diamond Jubilee, which featured a more layered, three-dimensional version of the Diamond Nine formation (Deep Diamond, see Glossary), as devised by Jim Turner. The year also saw fly-pasts over Belfast, Cardiff and Edinburgh and over the Olympic Stadium in London (with the Chief of the Air Staff in the back seat of one of the Hawks) to mark the opening of the 2012 Olympic Games in London, as seen by a global television audience in excess of one billion.

The two accidents of 2011 had been covered extensively in the press, and the team was overwhelmed by the welcome it received during the display season in 2012, as Martin Pert, then flying as Red 2, remembers: 'The response from the public was very affectionate, and there was a real outpouring of grief about what the Red Arrows mean to the country. Everywhere we went, the reception when we were flying and when we landed at airports was fantastic and there would often be hundreds of people watching us land. It was one of the busiest times for people coming out to greet us and, for us, a really poignant year in that we were able to carry on, with the trust of our senior command and real warmth from the public.'

Training for the subsequent 2013 season was mostly about getting the team back to a permanent nine-ship formation, despite questions from some quarters as to whether the team should remain a seven-ship, as part of an overall defence cutting process. Nonetheless, with nine pilots on board and more serviceable aircraft at the Reds' disposal, training went ahead. July 2013 marked the 4,500th Red Arrows display at RAF Waddington International

Air Show; and earlier that same year, Formula 1 drivers Lewis Hamilton and David Coulthard dropped in to fly with the Reds, while BBC *Blue Peter* presenter Helen Skelton also visited the team to raise money for Comic Relief's Red Nose Day. Skelton first flew to RAF Valley with Red 10, Mike Ling, for a visit to the simulator, before returning to RAF Cranwell to fly a couple of trips with then Red 1, Jim Turner, looping and rolling as an eight-ship formation. 'She was fantastic,' remembers Jim. 'Generally speaking, women are better passengers in Red Arrows – I don't know why, they've got a better stomach for it. The female *Blue Peter* presenters in the past have had few problems, whereas quite a few of the male presenters are airborne for ten minutes before they are throwing up and we have to land.'

The fiftieth display season for the Red Arrows in 2014 saw the commission of a special one-off tailfin design to recognise the anniversary, plus a series of celebrations – the Red Arrows being the main feature and theme of air displays in the UK and beyond. With the public profile of the Reds riding higher than ever, various TV and radio documentaries and magazines featured the team. On 10 September 2014 it carried out a special fly-past over the Olympic Stadium to mark the opening ceremony of the Invictus Games – the team was scheduled to arrive at 19.02 and fifteen seconds and, in true Red Arrows fashion, performed the fly-past bang on time, to the absolute second.

I returned as Red Arrows team leader for the 2015 season, a year in which a fresh new paint scheme, a flowing Union Flag, was applied to the tailfin of the jets; this remains the biggest-ever change to the Red Arrows' livery.

After a busy domestic season in 2016, we embarked on our largest overseas tour in a decade: a nine-week deployment to the Asia-Pacific and Middle East regions. We visited seventeen countries, met many challenges along the way and displayed in China for the first time in the squadron's history. My last year as team leader in 2017 involved another tour to the Middle East and, for the first time in twenty years, to Pakistan; and we were invited to display in Helsinki in June of that year at an event that proved to be the best-attended spectator event of all time in Finland.

Squadron Leader Martin Pert took over as Red 1 in 2018, a year that saw the centenary of the RAF, with the Red Arrows spearheading various celebrations, including providing a finale to a fly-past of more than 100 aircraft over central London in July of that year. Earlier in the year, on 20 March 2018, tragedy would strike RAFAT when Corporal Jonathan Bayliss, Circus 3 and an experienced engineer in the team, was killed shortly after take-off from RAF Valley. The pilot, Flt Lt Stark, ejected from the aircraft but Jonathan, flying in the Hawk's rear seat, died after the jet hit the ground. Mike Ling was brought back in as Red 3, with the season marking his tenth year with the team (three years as Reds 3, 7 and 6, six years as Red 10). The year of 2019 saw the team carry out its largest-ever tour of Canada and the United States, spanning eleven weeks and performing for tens of millions of people; and the team was honoured to fly with two astronauts in the back seats: Canadian Chris Hadfield in Toronto, and Britain's Tim Peake.

In 2020 the Covid-19 pandemic led to the cancellation of all UK mainland air shows – a first for the team – although the Reds carried out displays in Finland (marking an anniversary of the Hawk's use by both the Finnish Air Force and the Red Arrows), a display in Guernsey and a handful of high-profile fly-pasts to mark national anniversaries, including those for VE and VJ Day 75. The team also joined its French counterparts in fly-pasts over Paris and London to mark the eightieth anniversary of a speech by Charles de Gaulle to Occupied France from London. During 2020 the Red Arrows also carried out more than 150 online events, podcasts and video recordings, reaching out to more followers of the team across the globe than ever before. To ensure the cognitive flying skills of the pilots were not lost, training continued at RAF Scampton with flying positions 'frozen', so that Martin Pert could hand over to new Red 1, Squadron Leader Tom Bould, a team that was ready and able to take on the challenges and new adventures ahead.

5

The Pursuit of Excellence

A FEW MONTHS AFTER BEING selected for the team, I embarked upon my first Red Arrows display, this time as a passenger of Red 3's Hawk, so that I could learn more about my future role in the team. Heading towards the Mediterranean resort of Monaco, we flew in perfect Battle formation, nine red Hawks sprinting through the sky in a V-shape, each just twelve feet apart as we crested the ridge, diving down into Monaco harbour, before pulling up into a graceful loop to start the display. It was a glorious example of precision flying – all nine aircraft working as one, showcasing everything that was expected of Red Arrow pilots – yet all I could think was, 'I'm not sure I'm going to crack this.'

Now, of course, I realise that those feelings of self-doubt are common to many new pilots confronted by the seemingly impossible standards demanded by the Red Arrows. An aerobatic team with a global reputation, they strive for excellence (or '*Éclat*', as our motto states): a collective vision and aim that are, unsurprisingly, rather daunting for their new recruits. However, year in, year out, new team members manage to achieve that benchmark of excellence, or at least something very close to it, through a well-worn and graduated training programme designed to build up flying skills and increase their confidence and mental resilience. Repetition is key, with the pilots flying sortie after sortie in a gruelling regime that sees them gradually build up their self-belief and overcome their initial misgivings. This teaching mechanism is nested in the team's cultural approach to training and task achievement – one that promotes an environment within which honesty, integrity and humility are expected to be shown at all times. Objectivity and self-critical analysis are key.

All the pilots who join the team are accomplished fast-jet pilots, many of them have previously worked as instructors, and all have operational experience. However, starting at the Red Arrows can sometimes feel as if you're learning to fly all over again. Soaring through the skies in close formation – rolling, looping, holding your position while inverted, and manoeuvring with eight other aircraft in total synchronicity – requires a level of precision flying that is far beyond anything seen in operational flying. To former Red Arrows leader Jim Turner, who joined the team in 2004, the demands of the training came as something of a surprise: 'I was a very experienced Jaguar pilot. I'd been on operations, I was at the peak of flying that aeroplane, I couldn't get more qualifications. Then I started flying with the Reds and felt like a beginner again, a common feeling among new pilots.'

Former Red 1 and experienced fast-jet pilot Ian Dick was similarly confounded by what was expected of him in 1968: 'As an instructor, I was used to flying in formation, but not with more than four aircraft. I could not comprehend how nine pilots could fly in the formation shapes and manoeuvres that made up the Red Arrows display. It was a complete mystery to me: I just didn't know how they did it – let alone that I could fly like them. It says a lot for Ray Hanna, and the team I joined, that any reservations or under-confidence I had were quickly put to rest by the training regime I was put through: simple "building blocks" of formation flying that allayed all my trepidation and enabled me to fly as number five in my first year on the team.'

The training regime today still revolves around that 'building block' approach and is as intensive now as it was in the 1960s. Little time is wasted once pilots have received the life-changing news that they are on the team, as they are often required only a couple of months later to shadow pilots during the tail-end of the season. New pilots must also reacquaint themselves with the Hawk T1, an aircraft that many front-line pilots will not have flown since training. Usually starting at the beginning of August, new recruits would head to a Hawk training squadron to undertake six or seven flights and about seven simulator sorties, until they are 'Hawk-current' again – from 2022 this training will be conducted solely on the Red Arrows. The training essentially reintroduces pilots to their raw flying ability on what is a very simple aircraft to fly, but one that doesn't give the same level of protection in the way more modern flight systems do.

Pilots who have experienced flying fourth- or fifth-generation jet fighters, like the Typhoon or the F-35, will be used to a platform that is all about the screens in front of them, the radar, the weapons management and how they are operating the overall system. In the Hawk T1 there are no computers, other than a 1980s-like satnav bolted on, and so the real-time feedback from the aircraft is very important. When you put the Hawk into a manoeuvre and then take it right to the edge of its performance envelope, the jet responds beautifully, giving you this little burble that indicates it's at its maximum lift, and if you want anything else from it, you're not going to get it. It's very different in a modern fighter, where it's all fly-by-wire and computer-regulated, and you're able to put the stick straight back or straight forward and the jet will only give you what the performance element will allow. In the Hawk, however, if you're not attuned to the feedback from the jet, you can over-perform the aircraft and get yourself into trouble if you do so.

Once Hawk-current, new pilots join the team around mid-September, during which time they sit in the back of all briefings and fly with various team members because, as yet, they don't know which position in the formation they will be taking over.

They are also given a large file known as the Red Arrows' *Display Directive* that contains masses of SOPs, which they are required to go through so that they can digest all the intricacies of the various formation procedures. After a couple of weeks the team then throws a dinner for the pilots and their partners, and Red 1 will announce who is going into which position the following season. It's a significant moment for the whole team, especially for the new recruits, who will for the first time have a Red number – 2, 3, 4 or 5 – by which they will be known for the whole year. New pilots will then fly with their respective numbers, my own first trip being to Monaco, Malta and returning to Blackpool in the back of Red 3. Knowing your number serves to sharpen your focus even more, as you confront the huge learning curve ahead of you; it is one that, in theory, should be within your capability, if the rigorous selection process has done its job properly.

Once the previous display season is fully complete, by about early October, new pilots start two weeks of intensive training with Red 1 only. This is when they start to build a bond with their Red 1 and learn the core fundamentals of flying with the Red Arrows, plus the basics of close-formation aerobatics. Over three thirty-minute sorties a day, the new pilots perform loop after loop (after loop), before moving on to rolls, and that's pretty much all you focus on for two weeks. They start flying at 3,000 feet and, once Red 1 believes they can safely come down to a lower height (known as the 'step-down' process), they then fly at 1,000 feet and finally, after two or three sorties, at 500 feet.

These early training sessions start to build cognitive patterns so that when pilots pull up into a loop, they know what to do with the power of the aircraft. You learn where you put the throttle as you reduce speed at the top of a loop, and how you manoeuvre the control column – the pilot using very slight movements of their hands to adjust the steering; and if you keep doing all that repeatedly, it builds up a strong memory database of manoeuvres in the brain, which is a hugely important part of the training cycle. By breaking down those rolls and loops, you learn a lot about self-awareness, about what's working in the air and what's not working, and this is then confirmed by going over a visual recording of the sortie in the debrief afterwards. If you come out of position in the formation, you learn how to get back in position safely. And within those first two to three weeks you rapidly overcome that nervous psychological barrier of flying aerobatically at close quarters, around six to twelve feet away from other aircraft. During briefings Red 1 and the team repeatedly go over emergency procedures and 'escape manoeuvres', so that you know you can safely separate from the formation, should you need to.

Former flying instructor Kirsty Murphy found the first two weeks of training intense and was immediately struck by how different it was from her former role in the Royal Air Force: 'There were three of us, and my instructor brain was interested in how we were going to build up the training. We went straight in, and training ramps up quickly. It's all about repetition: loop, loop, loop, right turn, right turn, et cetera. I loved it. I was used to any trip in the Tornado taking about three hours of planning and preparation before you got in the aircraft, whereas in the Reds you briefed, walked down the stairs, got in the aircraft and off you went. It was so refreshing – it was just about flying, just about being the best pilot you could possibly be.'

ASSEMBLING THE TEAM

The next major phase of training is when Reds 4 and 5 join the front formation, with Reds 1, 2 and 3, to make up the section known as Enid. With Reds 4 and 5, who are often second-year pilots and fly on the outside of the formation, Red 1 will introduce some more technically difficult manoeuvres, such as the Quarter Clover, a manoeuvre that involves pulling up into a loop and twisting the formation by ninety degrees while pointing straight up, before pulling down to complete the vertical element of the manoeuvre, perpendicular to where you started. To fly in formation, pilots must align two chosen reference points on the front and back of the leader's Hawk – for instance, his wing-tip align-ing perfectly with the very front of his aircraft, or a specific bolt on the fuselage lining up with the tailplane section. Different for-mation shapes require different reference points, as all aircraft reposition themselves relative to the leader's aircraft – which per-haps sounds pretty achievable, but if you're upside-down, flying at 400mph in turbulent air, maintaining those reference points is more than tricky. During a training flight, Red 1 might also allow for something called 'Open up and relax' – thirty to sixty seconds of reprieve from aerobatic flying by remaining straight and level. It may not sound long, but it offers a chance for pilots to breathe and collect their thoughts, and enables the more experienced members of the formation to offer some debrief comments to the new pilots. It's a good example of the intensity of training, where every second is utilised to extract the maximum training benefit possible, with-out degrading the performance curve of the pilots.

Once Reds 1 to 5 are flying together, the team leader will grad-ually introduce the pattern of manoeuvres and shapes that he intends to fly for the new display season. Then the team will focus on the flow of manoeuvres, refining the transitions between them: you might arrive in a loop and then twist, go into a roll, followed by a reverse left, as the leader builds that particular choreography into your brain. As you rehearse every manoeuvre, your confi-dence will grow and you will become aware that everything you do affects everyone in the team, and so you will strive to do your tiny bit in each sequence exactly on time. Having got over any initial feelings of trepidation, tension is now more likely to build at debriefs, especially when the same minor faults are occurring repeatedly. You can often sense the frustration and self-induced pressure in the air amongst pilots who are trying to fix those stub-born minor issues!

The briefing and debriefing process, lasting approximately thirty minutes before and after every sortie, is an essential element of the training cycle. In briefings, Red 1 will outline the 'domestic' factors, such as weather and jet availability, amongst other things; and will comprehensively run through the sortie, from taxi-out and take-off, to breaking and re-joining procedures, which manoeuvres will be flown, and how – along with Red 1's corresponding radio calls – and when they will be pulling G. They will also discuss at least one emergency procedure; for example, Red 1 might ask a pilot to talk his or her way through how they would handle an escape manoeuvre, should an aircraft need to pull out of formation, or in the event of RT (radio) failure. This is a critical process, which encourages the team to visualise the task ahead and primes the brain to be ready to deal with any unforeseen circumstances.

Performing at your best necessitates mastering ways to deal safely with the unexpected. Our method is mostly verbal and dia-grammatic: Red 1 might move magnetic jets on a whiteboard to clarify formation references, while the team pilots literally count out loud the seconds it takes to transition between two shapes, to make sure they step out of the briefing room with their timings perfectly in sync with the rest of the team. Different aerobatic teams around the world have their own different methods. The US Air Force Thunderbirds (a magnificent team of F-16 fighters) spend a lot longer than we do in their briefing cycle and incor-porate more factors into the brief, including elements such as the timing of their commentator and the music that they will be flying to. The French team, the wonderful Patrouille de France, briefs in a more physical sense than we do (at times), by closing their eyes and shifting around in their seats to visualise their sequence. The different approaches are fascinating to observe, and no one method is better or worse – it's whatever works for the team and produces the best execution of the sortie that counts.

Every sortie is followed by a thorough debrief in which the team, including Red 10 – the team's safety supervisor and display commentator, who often flies the Red Arrows' photographers – and other senior officers, assess video footage, sometimes frame-by-frame. Each pilot is encouraged to call out his or her own mistakes as they scrutinise their position in the formation, while the leader and other team members highlight any other issues that they spot.

In looking at the footage, if a pilot judges that their aircraft is too close or far away from another aircraft, they will say it is tight or wide; too far forward or backward in the formation, and it's short or long; too high or too low, and they will say it's shallow or deep respectively. Precision-formation flying is a complicated business! Any query or fault, however small, will be raised and discussed, and team members must also draw attention to any difficulties or potential hazards they may have encountered during the sortie. Briefings are concise, entirely transparent and are run with absolute efficiency. A one-word answer will suffice and, to the untrained ear, much of the briefing is hard to follow, but the team covers a lot of ground quickly, with pilots focused on what they need to improve before they prepare for their next sortie.

Whenever possible, we open our doors to visitors, so that they can sit right behind the pilots and watch this human dynamic unfold. Often this part of our training cycle gains the most interest from our visitors, especially those from other military units or anyone in the general public who has an interest in strong team dynamics. For our team, de-personalising the debrief is fundamental to honest feedback and constructive criticism. When a pilot flags the mistake of another person, it is with reference to achieving the collective task, not as a personal criticism of the individual in question. To help achieve this, we talk to each other by number, not by name. To the onlooker, this may seem very odd and rather robotic – watching two people identify themselves as the number two or seven simply seems weird – but for us it creates the right psychological environment where, in the professionalism of the debrief, we reference each other and ourselves only by our role within the team. As soon as the debrief is finished, we are back to first-name terms or, more likely, to our familiar nicknames.

By the Tuesday of a typical training week, Red Arrows pilots will probably have taken off and landed more times than they would have done in a month on the front line. On a standard training day the team will fly three sorties (weather permitting), beginning with a first briefing at 8 a.m., followed by a sortie, debrief, and so on, in a process that runs from Monday to Friday right up to Christmas and then on to Easter. It's an intensive schedule and very different to life on the front line, where you might spend four hours planning a mission, six hours in the aircraft, three hours in the debrief and then perhaps not fly for another six days. The frequency and regularity of our sorties give Red Arrows pilots a truly unique opportunity to build upon their raw flying ability. This is one of the best parts of the job.

New pilots must also train using the smoke system, which they will begin to use the moment they start training with the team. The Red Arrows are famous for their colourful vapour trails, which provide real visual impact and facilitate many of the manoeuvres they do, such as the Heart and Carousel (see Glossary). The vapour is also vital for safety, allowing Red 1 to judge wind speed and direction and the pilots to see each other during head-to-head manoeuvres, or when different sections of the formation are several miles apart. For the bulk of training the basic white vapour colour is used, with the iconic red and blue colours added only for the last five weeks of Springhawk (the final stage of pre-season training) and for the display season.

The Hawks are fitted with several buttons and switches on the control column, with three buttons that regulate turning on and off the smoke. It's a very simple system, especially when compared to something like a Typhoon, a more complex aircraft, where you have multiple functions for both your right and left hands on the stick, ranging from selecting autopilot and weapons, to navigation and air-to-air refuelling. Despite this, you'll see seasoned fighter pilots, who just a few weeks earlier were flying fourth- or fifth-generation jets, struggle to operate those three simple buttons correctly, leaving smoke trailing when it should be off, or entering a loop with no smoke at all. All this overtly demonstrates the mental capacity and meticulous concentration that are required to fly safely in close formation.

When flying, each aircraft can produce smoke for a maximum of seven minutes, with five minutes of white smoke and one minute each of red and blue smoke. For this reason, each display has a carefully worked out 'smoke plot' to ensure that no one runs out of smoke when Red 1 makes the call 'Smoke on, go!' and that we trail the red, white and blue colours in that particular order. For high-profile fly-pasts, such as the Queen's Birthday, Red Arrows SOPs state that if you're on the right-hand side of the formation (4, 6 and 8), you smoke red; if you're in the middle (1, 2 and 3), you smoke white; and if you're on the left (5, 7 and 9), you smoke blue. It's pretty easy: 'red on the right'! However, when things aren't going quite to plan, you might see red, blue, red; or even purple (red and blue together), which might, if we're quite far into a training, incur a large personal fine. Any mix-up in a national event is deemed absolutely unacceptable. A £10 fine is the minimum levy for such an embarrassing error, but the feeling of letting down the rest of the team is a far greater burden to endure.

By Christmas, Reds 8 and 9 might join Enid (Reds 1 to 5, the front and outer parts of the formation) so that the team can fly as seven aircraft, while the synchro pair, Reds 6 and 7, continue working on their demanding manoeuvres separately. Flying as a seven-ship formation, the team will have covered much of the choreography and flow for the first half of each display, which consists of synchronised aerobatics by the whole team, plus around half of the more dynamic second half, manoeuvres that require the step-down process (see Glossary) and so carry a lot more training burden.

In building up the choreography, the team will work through each manoeuvre until they are consistently flying it safely, before moving on to the next one. One of the more difficult manoeuvres that we fly is the Rollback, which sees a pair of jets take turns to pull up out of formation, roll 360 degrees in a tight barrel roll, then slot back perfectly on to the outer edges of the formation. Get the technique wrong and there's a risk of being upside-down very close to the ground and not having enough height to recover the aircraft. Three pilots – Stephen Noble flying with Dennis Hazell, in 1978, and Neil MacLachlan, in 1988 – lost their lives while practising

Rollbacks and, as a result, the manoeuvre has evolved into something of a hurdle for team members to overcome. To ensure it is carried out safely, new pilots will practise with more experienced pilots in the back seat at a minimum of 1,000 feet above the ground. They will practise doing one barrel roll at a time, and only when this is deemed safe will they transition to lower heights and then do the manoeuvre synchronised as a pair. For Red 1, this involves leading the formation up and down the runway, but offers the necessary training mechanism to keep this visually impressive manoeuvre in the display.

PULLING GS

While the rest of the formation are working on building up those manoeuvres and shapes, the synchro pair will have been rehearsing their low-level opposition passes, gradually and safely adding in extra manoeuvres. Both are experienced Red Arrows team pilots, and when Red 7, the newbie of the pair, comes to the end of the season and prepares to step up to Red 6, the synchro leader, it's his or her decision who will be their Red 7, the synchro protégé, for the next year. It's such a close partnership, requiring an extraordinary level of trust and collaboration, that the decision is Red 6's alone, and no one else will influence that. The new Red 7 then needs to learn to work in tandem with Red 6, performing synchronised opposition manoeuvres, rolls and intense G-force (a measure of acceleration, felt due to the force of gravity) turns at just 100 feet above the ground. The first skill to learn is how to manoeuvre the aircraft safely at that low level, which in itself requires a dedicated training package of fourteen sorties.

The aim is to reassure everyone, and not least the new synchro pilot, that they are safe to operate at heights as low as 100 feet above ground level. This capability is not specifically unique to the Red Arrows, because other Royal Air Force pilots undergo specialised training called 'operational low flying', which is necessary for tactical reasons. However, there is a nuance attached to the balance of focus in the Red Arrows. We want our synchro pair to be comfortable and safe operating in that low-level environment, but never to feel too familiar that they switch off. Instead we require the synchro pair to visually judge their height as they run in for opposition passes with each other. This means they can't simply stare inside the cockpit at their altimeter, but need to be looking at each other as well.

Flying at low heights of between 100 and 300 feet makes it much easier to consistently judge your height over the ground, no matter the display location, which in turn makes it easier to judge visually an accurate pass height and set a safe miss-distance with your synchro partner. The main reason is that when you're flying at 300–400mph over the ground at these heights, you get a very real sense of the ground rushing past you. This sensation, called 'ground rush', stimulates all the temporal sensors in the body and brain. From a professional point of view, this causes extremely heightened levels of brain activation, making sure that every single element of your mind is focused on the task. From a personal perspective, those first few sorties as part of the synchro pair were some of the most exhilarating and energising of my career. After every debrief I had a huge smile on my face, and I loved the huge rushes of adrenaline. Perhaps unique to those who have flown in the synchro pair, there is a small sense of disbelief that you have been so lucky to end up in this particular job, being paid to safely and skilfully operate your aircraft in such a way that it creates a significant part of the exhilarating wow-factor associated with the team.

Mike Ling, who was Red 7 in 2008 and Red 6 (my synchro leader) in 2010, remembers how different it was taking on the more dynamic synchro moves: 'It's a totally different kind of flying – it's low, it's fast, energetic and you're managing a few extra risks. When I trained, we began by flying off on our own, flying at low level, pulling lots of G – now they work more as a pair from the outset. It's pretty exhausting the first couple of weeks, but you get accustomed to it.'

Ian Dick (Reds 7 and 6 in 1969 and 1970) remembers flying

with Henry Prince, who was Red 7 in the very first Red Arrows team, and synchro leader the following year: 'I was lucky enough to fly with him on a practice display. What an eye-opener that was, and the memories of the way he and Ernie Jones flew the cross-over manoeuvres, including the famous "Roulette", are as clear today as then. I was amazed by their flying: it was not in any Pilots' Notes that I had read! It was all done in such a relaxed, professional way that made it look easy.'

Once you have established that you're safe and ready on your own, it's time to move on to the next stage – to start pointing your aircraft at another aircraft at closure speeds of up to 800mph.

During winter training the synchro pair will fly three sorties a day, along with Reds 8 and 9, and altogether they make up the rear section of the formation. They fly in rotation with the front five aircraft of Enid, and so when one section is airborne, the other might be taxiing out and readying to take off once Enid has landed. To the aviation enthusiast, this means a heavenly conveyor belt of aerobatic training all day long. Reds 8 and 9 have a flexible and dynamic job of their own, especially in the second half of a display – one minute they are pushing their aircraft to the corner of its performance envelope with the rear section, and the next they are re-joining Enid to conduct more low-tempo, highly graceful formation manoeuvres like the Vertical Break (see Glossary).

While the synchro pairs often relish the opportunity to work on their own, they are fully supervised by Red 10 on the ground – most recently Squadron Leader Adam Collins – who liaises with them and the rear section via the radio. His primary role is to check on safety, and he can also give Red 6 a 'hot debrief' (in-the-moment feedback) as they repeat manoeuvres during single flights, all of which will be recorded and used in the office debrief later.

For the synchro pair, much of winter training is taken up with flying opposition passes, which (like everything else) is done in an integrated way: in the first pass they do they will be quite far away from each other, and then they will slowly bring down the separation distance to a minimum of 100 feet laterally between the two aircraft. Once they have performed that consistently and safely, they will add more complex moves, perhaps a four-point aileron roll – when an aircraft does a full 360-degree rotation – crisply hesitating every ninety degrees. Another more complex move for the synchro pair is the Vice-Versa manoeuvre, which sees Red 6 flying inverted as he heads towards Red 7, the pair rotating as they pass each other so that Red 7 becomes the one flying inverted, with Red 6 now the right way up. The pair continues practising each pass and manoeuvre three times a day, finessing and polishing until they have it down pat.

The synchro pair and rear section perform the most physically demanding manoeuvres, experiencing high levels of G-force. All fast-jet pilots in the RAF receive regular G-training, attending a forty-eight-hour course where they spend most of the time covering many aero-medical topics before they are spun around in a centrifuge facility that accelerates pilots up to 9G. This mandatory training ensures that pilots are sufficiently aware of the physiological challenges inherent in high-G manoeuvring, whether it's in a Typhoon, an F-35 or a Hawk. Not only are new Red Arrows pilots up to date on this training, but a modified version of the centrifuge course is also given to the Circus engineers. Although they never fly in a public display, during the display season they are exposed to more airborne time than some pilots in other areas of the RAF.

Like all jet pilots, Red Arrows wear anti-G trousers, which inflate with pressurised air to stop blood pooling in the lower legs, keeping it in the upper body and brain to prevent loss of consciousness. Five to ten seconds of 4–5G can lead to tunnel vision and loss of consciousness, but the anti-G trousers can facilitate resistance up to 6 or 7G. The synchro pair, when performing an aggressive manoeuvre such as the Carousel, might experience 6–8G, which effectively causes the head and body to weigh six to eight times more than normal.

Before pulling any G, pilots are trained to tense their buttocks, leg muscles and calves and to take a breath before any manoeuvre, raising the heart's output to further prevent blood-pooling. Once you're under G, you continue a special form of breathing and you do this every time you perform a manoeuvre. Expected G-force pressures are flagged at every briefing, and if the team has had a five-day break, we'll start training with low-G manoeuvres in a 'G warm-up', just to get the capillaries going again and to build up G-tolerance. The way the body deals with G-force can be erratic: some days you might be a bit of a 'G-monster', able to withstand high Gs without any problem, whereas on other days, when you unknowingly have a virus or aren't fully hydrated, you might be far less G-tolerant. Losing consciousness due to G-force is a major hazard to all fast-jet pilots, so we ensure that a declaration is made before every sortie in what's called the 'out brief', which is a final check that everything is in order prior to flying. The first confirmatory check in the out brief is a question put to all the team pilots: 'Is everyone fit to fly?' This is the last chance to declare whether you feel physically and mentally ready for the task ahead. If you say no, there is zero blame attached, and although it is extremely rare to declare it so late in the preparation phase for flying, having the moral courage to say this is praised rather than looked down upon. Nonetheless, the G-force risk remains real, and in 2011 fellow team member and a very close friend of mine, Jon Egging, lost his life as a result of the G-environment while performing a routine landing manoeuvre – a risk to all fast-jet pilots, not just a Red Arrow.

Pilots and Circus engineers also receive hypoxia (oxygen-deficiency) training. At 10,000 feet and above, as the atmospheric pressure reduces, so does the available oxygen content in the air. As a result, fast-jet pilots wear oxygen masks at all times to ensure that the correct oxygen levels are given, because lack of oxygen can cause a range of symptoms that are often difficult to spot. As the only way to recognise the early signs of hypoxia is to experience them, training involves sitting in a hypobaric chamber, where personnel might be asked to perform simple childlike tasks, such as fitting shapes into the corresponding slots on a coloured ball. As the oxygen is reduced, they may struggle with their task, become confused or simply feel a little punch-drunk. Everyone reacts differently – some become relaxed and euphoric, while others may become frustrated and impatient. If pilots learn how to spot these early signs, they can take action quickly. When the team transits across the world we fly at high altitude amongst commercial airliners, in what's called the 'airway structure', a mass network of specified routes that ensures everyone gets from A to B safely. For this reason, everyone on the team has to know what their hypoxia symptoms are and ensure that an emergency situation is dealt with properly, by immediately descending to lower altitudes and activating the emergency oxygen system that provides 100 per cent oxygen for a limited time – enough for a pilot to get back into a safe operating altitude.

THE FIRST NINE-SHIP DAY

Once the synchro pair and rear section have finalised their dynamic manoeuvres, the aim is to get all nine aircraft together in around early March, in the latter stages of winter training. Prior to that happening, Reds 1 and 6, the leaders of Enid and the rear section respectively, must fly timing runs to ensure that when their sections split from each other in the second half of the display, their choreography meshes well and what seems to work on a whiteboard actually works in the air. Once Red 1 and Enid have finished a manoeuvre, Red 6 and the rear section need to be in exactly the right location to fly in and perform their sequence without leaving a large gap of empty sky in front of the spectators, thus ensuring that displays are constant, dynamic and flowing.

When each section is on the display site, it effectively owns the forty-five to seventy seconds of audience viewership, known as 'the contract'. We call the changeover between the section leaders, Reds 1 and 6, the 'contract handover'. For air safety, the calls we use, such as 'Smoke off, go', aren't just radio calls to tell the pilots when to turn their smoke off, but also indicate this important handover. When Red 1 calls for the smoke to be turned off at a certain

1. The 1932 RAF Pageant at Hendon. In the face of major interwar cuts, Air Marshal Sir Hugh Trenchard, 'Father of the RAF', saw the value of public displays for showcasing the skills and effectiveness of the RAF. (All images courtesy of UK MOD © Crown copyright unless specified)

2. A specially-built set-piece is bombed by pilots during the 1928 Hendon Air Display. In the interwar years, the RAF would stage mock battles, recreating famous war scenes as aerial theatre, to showcase their aerobatic flying skills.

3. The five distinctive yellow Folland Gnat T1s of the 'Yellowjacks' display team, pictured during a practice flight.

4. The yellow Folland Gnat T1s were resprayed in postbox red and The Red Arrows display team was formed. The team's name, inspired by the colour and arrow-like shape of the jets, is also a nod to the earlier Red Pelicans and the Black Arrows of 111 Squadron.

5. The Red Arrows pilots team photo, 1965.

6. The Red Arrows pilots and ground crew at RAF Wattisham in 1965, the home of the No. 1 Air Control Centre from 1967 to 1979.

7. Squadron Leader Ray Hanna, Red Arrows Team Leader, pictured in 1966, shortly after the Red Arrows first formed.

8. The Red Arrows Gnats over-flying Niagara Falls during the team's first ever tour of North America.

9. The Nine Hawk T1s pictured over the Pyramids at Giza, Egypt in 2003.

10. The Red Arrows fly over the Olympic Stadium in London during the Opening Ceremony for the 2012 Summer Olympic Games. (Courtesy of Getty © Elsa/Getty Images Sport)

11. The Red Arrows fly in mixed formation with the last airworthy Vulcan in the world, XH558, after displaying at the 2012 RIAT.

12. The Red Arrows fly in close formation with Concorde, for the Queen's Golden Jubilee Fly-past in 2002. Concorde was retired from service the following year.

13. The Red Arrows escort a British Airways Boeing 747, flying in mixed formation.

14. A beautiful Spitfire, of the Battle of Britain Memorial Flight, flies alongside Red Arrows jets as part of a mixed formation fly-past to support the opening day of the Farnborough International Air Show.

15. The Red Arrows fly in mixed formation with the Lancaster bomber of the Battle of Britain Memorial Flight.

16-17. The Red Arrows had the privilege of meeting the Duke and Duchess of Cambridge with Prince George in July 2016. The Duke and Duchess were visiting the Royal International Air Tattoo, Gloucestershire, where the Red Arrows were displaying. The Duke of Cambridge was given a tour of the Hawk T1. Prince George also got to sit in the world-famous jet.

18-19. Top and bottom left: Queen Elizabeth II looks on with delight as she watches a fly-past by the Red Arrows over Windsor Castle, to mark her Official Birthday on 12 June, 2021. (Courtesy of Getty © Max Mumby/Indigo/ Getty Images Entertainment)

20. Since their formation in 1964, the Red Arrows have put on head-turning displays for gathered crowds, air show attendees, foreign dignitaries and, of course, royalty. Princess Margaret is pictured following the flight of the Red Arrows during a display at RAF Oakington, near Cambridge in 1967. (Courtesy of Getty © Douglas Miller/Stringer/Hulton Royals Collection)

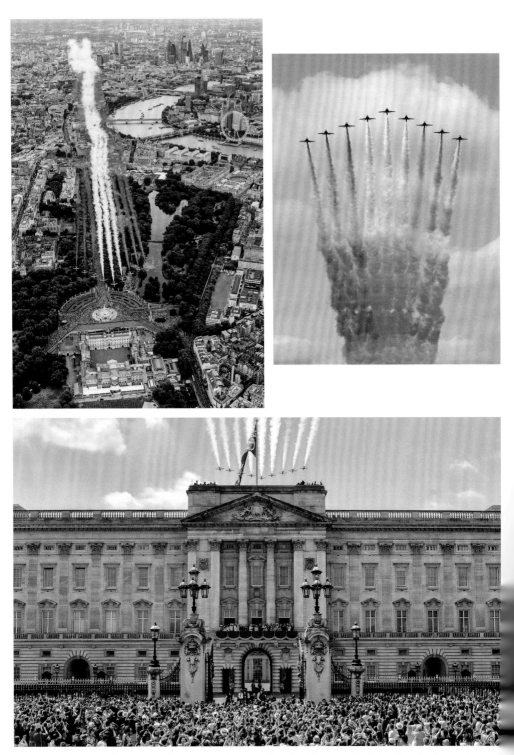

21-23. The Red Arrows are often a centrepiece of national celebrations, their iconic red, white and blue smoke trailing across the sky (top right). The British Royal Family and the gathered crowds watch the Red Arrows fly over Buckingham Palace during the RAF 100 Parade and fly-past in 2018 (top left) and the Trooping the Colour parade in 2017 (bottom). (Bottom image courtesy of Getty © James Devaney/WireImage)

point in the second half, this also means that Red 6 now owns the radio and can start directing the next manoeuvre. There are around seven or eight handovers of this nature in the second half of the display. At times you can hear Red 1 or Red 6 calling '1 late' or '3 early' as the contract is handed over, meaning that they are one or three seconds late or early. Accuracy is key and this information is vital, as the section not on show must now make sure their movements are subtly adjusted to ensure a perfectly timed contract handover the next time around. Adjusting the time you return into the display area is done by old-fashioned stopwatch timing – switching on and off a very analogue-style stopwatch that flips up on the instrument panel – and by using the compass rose to tweak geographical positioning. This technique doesn't rely on any modern-day GPS tracking or high-end navigational device (although we do now use GPS as an excellent back-up!) It's a proven methodology that leads directly back to the founding members of the team many years ago.

Once the timing runs and contract handovers are refined and finalised, the nine aircraft can finally get together. That first 'nine-ship day', when all nine pilots check in their numbers prior to take-off, is a significant one for the team, who will, for the first time, be able to bond as a formation in the air. Red 1 will be relieved to have reached that particular milestone, and new team pilots will be edging even closer to the dream of becoming an official Red Arrow. Thereafter much of the work involves bringing together and finessing all the various strands of the show. For the engineers, this is also a proud moment, as their unrelenting toil in maintaining the jets throughout the winter period comes to fruition. With such a personal bond to the aircraft, there is a huge lift for the whole team after 'first nine'.

With the synchro pair on board, the team can create the trademark Diamond Nine formation, with Reds 6 and 7 forming the central part of the formation, known as the 'stem'. This now enables Reds 2 and 3 to practise flying down the stem, a skill that is also required for such manoeuvres as the Concorde and the Swan, for which Reds 2 and 3 effectively have to reverse while doing a loop, so that they can get into the right formation position to set the shape. This is no mean feat for Reds 2 and 3, as they need to fly backwards without actually looking backwards, something that requires total awareness of their positioning in relation to the leader and synchro pair in the stem behind them. To conduct this manoeuvre safely requires a procedural approach – a tiny reduction of power, followed by a count, until a formation reference is seen that will prompt Reds 2 and 3 to make an infinitesimally small correction to their steering so that they slide smoothly back into position. Moving backwards into Swan is a formation change unlike anything I had ever been required to do in my previous life as a fast-jet pilot; it requires detailed SOPs and dedicated training, together with lots of practice. While Red 1 is supervising in the air, Red 10 is on the ground providing feedback when needed, checking for safety and contributing to the crucial debrief afterwards.

The third phase of training, employing all nine Hawks, usually involves two or three weeks' practice in the UK, before the whole squadron deploys overseas for Exercise Springhawk. Whether it's at RAF Akrotiri in Cyprus or at an airfield in Greece, the consistent blue skies of the Mediterranean enable the team to train intensively and to finesse the choreography, the aim being to have a safe display that is as perfect as we can possibly get, and one that meets the exacting standards and reputation of the team. Back in the 1960s and 1970s, the Red Arrows conducted all their training in the UK, despite regular requests from team leaders to have some time training overseas. Nonetheless, they made do with some intense training weekends whenever the weather was decent enough, packing in four sorties a day, during which they could make real progress. From 1980 onwards, when the Red Arrows were able to train overseas, those final weeks of flying in good weather enabled the team to further polish their display. It also gave them an opportunity to focus on any problem elements in displays, which, being curtailed by the unpredictable UK weather, had so far proved difficult to solve, as former Red 1 Jim Turner, then a new team member, discovered in 2005: 'We were out in Cyprus, just a few weeks away from display approval, thinking: "We're never going to be able to do this." But because you're flying three times a day in good weather, you have the momentum to get you through it, and suddenly you realise you're doing it, and that's why the team trains so hard, three times a day – it's the only way you're going to get up to standard.'

The whole of the Red Arrows team, including the engineers, the Circus (who now begin to fly with their respective aircraft), pilots, technicians and support staff are all deployed overseas. The display season relies on the smooth running of the entire squadron, and the Springhawk training enables the team to concentrate on building that cohesion, away from the usual distractions back at home. It's also vital to take the whole team away to a new geographical location, where not only are they assured good weather for training, but they can also fly over mountainous terrain and use different runways; and Reds 6, 7, 8 and 9, in particular, can practise their dynamic manoeuvres over the sea and the coastline. Flying over water can prove challenging for pilots as they don't have the usual ground reference points and the water can throw up all sorts of visual illusions. A haze over water, or flat and mirror-like conditions where the sea reflects the sky, makes the judging of the horizon and of height extremely difficult for pilots, particularly for the synchro pair, who are performing head-to-head manoeuvres as little as 100 feet over the water.

Pilots get used to flying at Scampton, but the team is required to fly across the world in vastly different locations and weather conditions. As Red 3, my first ever public show was in the northern tip of Jordan, ten miles from the Syrian border at Al-Mafraq. We flew there directly from Cyprus, had an ambassadorial engagement in Amman, then a display the following morning over desert-like terrain in thirty-plus degrees Celsius, all of which can reduce the aerodynamic and power performance of the aircraft. Our next show was five days later in Southend-on-Sea, Essex, where we experienced 20–30mph winds, rain and a low cloud base –

conditions that couldn't have been more different. The relentless and repetitive training, and choreography that enables us to safely fly different kinds of displays according to the varying conditions – a full looping display in Jordan; a flatter display with tight turns and shapes in Essex – builds on that all-important cognitive ability to alter our techniques of flying very slightly to suit a specific environment. This flexibility distinguishes the Red Arrows from other national formation aerobatic teams and is something of which we are very proud.

The months of training, and final weeks of further polishing during Springhawk, ultimately lead to the annual assessment known as the Public Display Authority (PDA), without which the team would not be able to perform their summer season of displays. Prior to the two-day PDA assessment, a one-star officer (Air Commodore) flies with the team. On his recommendation, the Air Vice-Marshal will then come out to visit the team for two days. The Air Vice-Marshal's aim is to assess whether the entire squadron is working well together and is fit and ready to be allowed to display to the public. The team will be assessed on a variety of displays, including the flat routines that we fly under low cloud, synchro-pair manoeuvres and even Red 10's commentary for the show. By the time of the PDA the team's performance is normally the best it will ever be – a standard that we'll strive to attain, and maintain, for the duration of the display season.

The Air Vice-Marshal will also scrutinise the synchronised 'see-off' performed by the Circus (see Glossary), as well as communication between pilots and engineers and the entire planning for the display season, including major engagements and high-profile campaigns, plus fatigue-management plans for team members; he will also assess the risk mitigations applied to all the display sites for the duration of the season. He will join the squadron for dinner and might sit next to a nineteen-year-old apprentice to get a real sense of the dynamic of the team and how well all the elements of the Red Arrows work together. By granting us the PDA, the Air Vice-Marshal – who is in effect the risk owner for the squadron – is saying to the Chief of Air Staff and the Secretary of State for Defence that the team is, in their belief, safe to do everything that is planned for the display period.

Gaining the PDA is a big day for the team, just as it was in the 1960s and 1970s when clearance was secured by the Commander-in-Chief of Flying Training Command. All those months of training have effectively been working towards this day, and once you've got that tick in the box by the Commander-in-Chief, the team adheres strictly to the version of the display – its full, rolling and flat choreography – that has been signed off. This is a principle that the team has worked to since its formation all those years ago, and to change the display, add new moves or revise it in any way would simply be too dangerous.

Another outcome of PDA clearance is that the team can swap their green training suits for the famous Red Arrows flying suits, which have been hanging in the pilots' lockers throughout Springhawk, serving as a visual reminder of what we are busy working towards. Putting on that red suit for the first time is a phenomenal feeling for any new pilot, as you are now a bona fide Red Arrow – something that you didn't dare feel worthy of before earning that suit. Even if you now break your leg walking towards your Hawk and are out for the rest of the season, you've made it and have met the standard. Similarly, and equally proudly, the support team switches to wearing their distinctive blue coveralls.

What came as a bit of a surprise, when I strapped myself into my aircraft wearing the red suit for the first time, was the way it reflected off the canopy, even with the lifejacket and anti-G trousers over the top. In 2009, as I flew from Cyprus to our first display in Jordan, the cockpit seemed incredibly bright, as flashes of red darted around me – as if I was now doused in some kind of magic – a strange experience that also surprised my fellow new pilot, Zane Sennett. As we stepped down from the aircraft in Amman and walked towards the debrief, we quietly commented to each other how unexpected it was to feel such childlike excitement over seeing our new red suits reflecting in the canopy: a subject that we were far too embarrassed to bring up with the rest of the team. While it may have, at first, felt as if we'd been anointed by some higher force, the effect of that red suit eventually seemed entirely normal after a couple of trips. Like everything in the Red Arrows, what seems extraordinary at first eventually becomes par for the course.

A BRIEF NOTE ON FINES

A more informal element of Red Arrows training is the team's fining system, which serves to motivate and bond the team, while injecting a bit of humour during what can be a tough and relentless schedule of training and displays. Known as the 'pigz' – after the 'Pigz Boards' found at most RAF squadrons – the fine system is Red Arrows-specific and is operated by the Executive Officer Red 8 or 9, who will mark down any minor flying indiscretions on their business-card-sized notepad and a small fine is given. At the end of the week they tally up and annouce who owes money to the pilots' fund. Fines relating to airborne operations only kick in after the team is safely flying as a nine-ship formation and, even then, often only once they start their final-stage Springhawk training.

When it comes to airborne and operational matters, the pigz is made up of an elaborate system of rules, as the team pushes itself to perfect and finesse its knife-edge routines and manoeuvres in the sky. Fines can kick in on a whole range of misdemeanours, from a half-second delay in the pre-flight 'check-in' to a £1 fine if the front wheel of a pilot's Hawk isn't lined up exactly, to the inch, on its parking spot. Once airborne, if you put on the smoke more than half a second too early or too late you might incur anything from a 20p to a 50p fine – all of which can add up during a show where there are typically about fifty 'Smoke on' and 'Smoke off' radio calls.

However, 'on the ground' misdemeanours are not exempt from fines, and in fact as soon as new pilots arrive at Scampton they are given the entire pigz set of rules, which they must read through and digest. Most of the ground indiscretions relate to timing, which dominates so much of what the Red Arrows do. If pilots are even a second late to weather briefings they will get a fine, generally charged at £1 a minute. If a pilot misses a deadline for a benign duty, or a male pilot turns up to work with an unshaven face, that's a £5 fine off the bat – a rare event, but no less excusable. Fines can also apply to whole sections of the formation, and there is often a bit of friendly competition between Enid and the rear section, just as there often is between senior and junior pilots in the RAF (although Red 1 generally supports the newer Red Arrows pilots in Enid, as they train and perform together).

One of the most lucrative activities for building fines relates to 'wheels', which in RAF parlance relates to any bike, car, truck, tuk-tuk or coach that carries the pilots from one place to another. If Enid and the rear section are in separate wheels, as a rule the rear section will end up with the better vehicle. With the Red Arrows, everything must be done to the second, so if wheels are used to get from the team's accommodation to the airfield, then the wheels must leave the front of the accommodation building – whether it's in Blackpool or Bangalore – exactly on the WHAM timing (see Glossary). If the vehicle is moving a second late, all of the vehicle's passengers take the fine. While we're careful not to end up with a *Wacky Races*-type affair, and the morning kerfuffle can often get some members of the team a little fired up, this does help to instil the importance of essential Red Arrows skills such as accuracy and punctuality.

On a Friday morning at the end of each week the Executive Officer will inform the team how much each member owes, at which point they have the option to go for 'double or quits'. Then either before the penultimate or last flight of the week, on the taxi-out before take-off, a pilot can call Red 1 and request a 'clear round'. Should he get the go-ahead from the boss, that pilot has the chance to clear all fines incurred, if he or she performs a sortie without any mistakes or just one (you're allowed one mulligan). Commit more than one mistake, however, and you risk doubling the amount you owe. Sections within the formation can also call for a clear round, and the same rules apply. The tradition is that, on the final trip of the PDA, the whole team aims for a clear round, with Red 1 calling on taxi-out, 'Red Arrows Clear Round'.

The fine system, under the watchful eye of Red 1, always works within the strict safety parameters of the Red Arrows and reflects the bonhomie of a team that adheres to the vision of collective goal-setting in the relentless pursuit of excellence, whether we're flying over a sometimes soggy Southend or the sun-baked hills of Monaco. A dazzling display, with spectators on the ground gasping in awe, and a big perfect zero on the Pigz Board: that really is the dream.

6

Red 1

THE ROYAL AIR FORCE has only one aerobatic team and to become its leader is a rarefied honour. You follow in the wake of previous team leaders, but once you're the 'boss' and effectively the only Red 1 in the country, you alone lead on the ground and in the air, with eight jets following your every move and radio call. It's a job that is unrelenting in its demands and responsibilities, one that team leaders live and breathe around the clock, with the pressures being heightened by the high-risk nature of the work.

My first three years as a team pilot with the Red Arrows were an incredible but tumultuous period, packed with the thrills of flying formation aerobatics as a new pilot, and then with the more dynamic moves as part of the synchro pair. In 2010 a mid-air collision with my synchro leader, Mike Ling, left him unable to continue the season. Miraculously I was able to keep going, with only minor scrapes. Late in August 2011 we lost fellow team member Jon Egging and, just weeks after I'd finished my first three-year tour with the team, we also lost Sean Cunningham, both to unrelated but tragic accidents. By the time I left the team in the late summer of 2011, although I had made it through physically unscathed, I had to process a great deal of mental turmoil and never thought for a second that I'd find myself back with the Red Arrows. But three years later, after a front-line ground tour working with the US Marine Corps in Afghanistan – which, in all honesty, felt like a break, after life with the Red Arrows – I got a tap on the shoulder: would I consider coming back as Red 1?

My time away from the Red Arrows had given me the space I needed to at least consider the proposition – or perhaps to forget some of the more stressful aspects of life with the team. After some procrastination, I eventually decided to throw my name into the hat, as one of four potential candidates for the job. Before I knew it, I learned that I had made it through and I was back walking through that familiar atrium of RAF Scampton, this time as the new Red 1, knowing that my portrait would join the ranks of the twenty-one former team leaders lining the walls. Going for the top job was undoubtedly one of best decisions I ever made – and I knew I would have regretted it, had I not gone for it – but I was under no illusions about the weight of responsibility and the many challenges that lay ahead.

The Red Arrows team works very much as a cohesive unit, with every pilot working towards that common goal of precision flying and agility, showcasing the best of the RAF. The experience of being up front, however, is markedly different from that of being Reds 2 to 9. As a team pilot, I used to love those seconds when you're lined up on the runway surrounded by a pack of red jets awaiting clearance to take off. All intra-formation chat would stop, the radios were largely silent and you were entirely focused on the job at hand. If we were headed to a display, my thoughts were chiefly that I didn't want to mess up the flight for the rest of the team and that, so long as I gave the best performance I was capable of, everything would be okay. Primed to go, I would feel a sense of elation before take-off, as I relished the prospect of what I was about to do.

As Red 1, however, those seconds before take-off felt entirely different, with my brain at times poring over thoughts of what lay ahead and the myriad different scenarios that might unfold in transit to a particular display. I was constantly thinking five to ten minutes ahead, and the tricky part was trying to bring my mind back to the 'now'. The stopwatch was always on that relentless countdown, and I was all too aware of the precise to-the-second timing that was essential if we were to arrive at a display site on time. If we came off this timeline, how would we fix it and where was my 'cut short' plan to make up any lost time? If a team member suffered a technical issue prior to take-off, where would I meet them, if not on the runway? And where was my next waypoint where we could stop and take stock? If a thunderstorm loomed, how would we safely fly around it and get to the display location? All these things bubble around your mind, and your brain whirs with calculations and thoughts, while you lead the formation. I suspect all team leaders have felt this way, at one time or another.

There were times when, sitting on the runway with the engine running, I almost forgot to breathe for what felt like a minute, because I was thinking about the minutiae of the task at hand. The more challenges we had had leading up to a display, the more this was likely to happen. For me, the displays or fly-pasts often felt like the easy part, because once we'd arrived in the display area we simply did what we'd spent all our time practising to do. In comparison, getting to the site on time often proved more logistically troublesome. At these stressful moments I remember saying to myself: 'Monty, just stop and breathe', and I'd take three deep breaths and try to reorganise my thoughts, before issuing the famous Red Arrows radio call for take-off: 'Reds rolling . . . now.'

RED 1 MATERIAL

To prepare for the rigours of the job, Red 1s go through a fairly brief period of training to develop the skills needed to lead a formation of aircraft. The role of Red 1 is open only to those who have flown with the team, as it requires a set of skills that is largely unique to past team pilots. Of all the team leaders, only Dennis

Hazell and Sir Peter Squire took up the mantle without having previously flown with the team, and while both were eminently accomplished pilots (the latter becoming Chief of Air Staff, no less), their tenures with the team were, for different reasons, short-lived, which proved to many people that previous team experience was an absolute must.

Many Red 1 pilots are former synchro-pair leaders, from more recent Red 1s, like me and the current incumbent Tom Bould, going right back to the likes of Ian Dick and Brian Hoskins in the early days of the Red Arrows. However, synchro-leader experience is by no means a prerequisite, and various leaders have flown in a number of different positions before becoming Red 1, including John Blackwell, who unusually had spent just one year in the team as Red 5; and more recently Martin Pert, who in his final year with the team flew as Red 8 – a challenging role that requires consider-able movement between front and back sections of the formation. The Red Arrows also prefer their leaders to have taken some time away from the team, as the three-year-long tours with the Red Arrows are intense and most Red 1s need to acquire further lead-ership experience on a 'command tour', whether on the front line or in a training unit. Only two leaders in Red Arrows history have been elevated to the coveted role from within an existing team: Bill Loverseed in 1971, who took over from Dennis Hazell after his accident, and Ben Murphy in 2010.

Like new pilots, team leaders must undergo training to re-acquaint themselves with the Hawk T1. Once 'Hawk-current' again, they then join the existing team as it finishes off the display season, shadowing the incumbent Red 1 for about four weeks, flying in their back seat to a few displays in the UK and abroad. During this time they must take in the technical elements of what it means to be Red 1, while observing the dynamics of the team and working out how they will lead the personalities who will remain. In Red 1's first year at least six pilots in the team will stay on for the next season, while two of the pilots will have reached the end of their three-year tour. In many ways, you feel like a real spare part during the shadowing period, skulking in the background while the current team goes about its business. However, observing the dynamic is a game-changer and, in my opinion, one of the most fascinating aspects of the handover, because you need to work out how your leadership style must adapt, as well as the different ways in which you might need to handle and support certain characters within the team. I'm confident that other leaders – in whatever walk of life or role – would agree that this 'shadowing' opportunity offers the best recipe for success in their first 100 days.

The new Red 1 definitively takes over as team leader at the annual end-of-season dinner, a black-tie military guest night that concludes the season for ex-team members and those who have directly supported the team during the year. After the formal dinner, at the stroke of midnight, the metaphorical Red 1 'baton' is handed over. For new Red 1s, this is the moment when they can properly step into the role and take over, and I remember being desperate to start, although I was no doubt daunted about what I was going to face. As with many difficult jobs, you never know how you'll fare until you are actually doing it – and leading the Red Arrows as Red 1 is no exception. Of all the positions in the team, this is the most challenging one: you have eight wingmen trained on your every move, you call the shots on the ground and in the air and, if anything goes wrong, you are responsible.

Dickie Duckett, when he took over in 1975, knew of other leaders who hadn't been able to rise to the challenge, so he keenly understood that being Red 1 'wasn't as easy as it looked'. Ian Dick, team leader from 1972, was similarly aware of the task ahead, but had legendary former team leader Ray Hanna on hand to give him some invaluable tips: 'I love formation flying, but I wasn't confident at all that I could lead nine aircraft. I had flown in lots of other positions, but leading was a complete mystery to me. I had no idea how to do it. I rang Ray and said, "Ray, how on earth do I lead the team?" He said, "Ian, imagine you're doing aerobatics in a big aeroplane, as if you're flying a Vulcan. When you manoeuvre the team, imagine flying a Vulcan; just don't roll it, because the left-hand wing-tip goes down and that makes it very difficult for the outside guys on the left. Whenever you turn, always pull up a little bit, so you roll or turn around the inside guy, whether he's on left or right." That was the best advice, and the only advice he gave me.'

FLYING BY EAR

After taking on board any nuggets of wisdom passed on from former leaders, a new Red 1 usually spends a good four to six weeks focusing on the essential flying techniques he or she will need to lead the formation. After flying in the back seat with the former Red 1 in training sorties at home base or in a handful of public displays, they will fly sortie after sortie on their own, performing aerobatics gradually at lower heights until they can do so safely at 500 feet above the ground. As with new pilots, Red 1 first starts with loops and rolls, manoeuvres that are fairly straightforward for most RAF pilots, but such moves take on an extra level of complexity when done while leading a tight formation of eight other aircraft. Key to flying a loop or roll as Red 1 is to be perfectly consistent (as perfect as you can make it) every time, as the other pilots in the team will become conditioned to their own techniques of manoeuvring their aircraft in response to your way of flying. Leaders must also avoid pushing the Hawk too hard, to its maximum aerodynamic capability, because the rest of the team will struggle to keep in formation and will 'fall off the wing'.

As an example, when entering a loop in a training sortie, Red 1 needs to get to a speed of around 400mph as smoothly as possible; once at the 'gate' or target starting speed, ready to enter the loop, Red 1 must radio-call to the rest of the team: '*Down* for the loop.' Once they have acknowledgement from the team, Red 1 calls out, 'Smoke on, go!' (they always practise flying with white smoke from the outset) '. . . and pulling *up*'. At the u of '*up*', Red 1 must start to pull the control column backwards in a graduated movement (we would say 'squeeze' it back), so that they hold three to three-and-a-half Gs (G-force) as they go up for the loop. One of the hardest things to learn, as Red 1, is that when you go into a loop and pull back on the stick, you not only have to smoothly apply pressure to get to the right G, but the stick also has to be perfectly central each time, so that when you pull up and get to the top of the loop, the wings on your aircraft are perfectly level to the exact degree – even two degrees off-kilter (or 'wing drop', as we call it) would have serious consequences for the performance of the manoeuvre: an unplanned twist in the loop is not what the team pilots are expecting and could compromise safety by leading to escapes from the formation.

Once you are at the top of the loop, Red 1 can't be too slow, otherwise the aircraft won't have enough lift and could 'buffet' and stall. If you fly too fast, that could make the loop bigger – you would be in danger of going into cloud at the top of the loop or, worse still, ending up below your minimum base height at the bottom of the loop and therefore find yourself perilously close to the ground. Pilots need to master the huge speed difference in the loop, going from 400mph at the bottom to 140–150mph at the top, finishing again at 400mph, or whichever speed is required for the next manoeuvre. It's these fundamental basics that Red 1 works on initially and continues to finesse throughout their two or three years as leader. And once they've got the basics of a loop down, there's a lot else they can do with it – whether that's twisting it halfway down to keep the formation more central to the crowd or incorporating a Quarter Clover, where you end up flying a ninety-degree loop, twisting on the way up – all of which takes a lot of practice and training.

Barrel rolls also take considerable practice; and rolling as a team – as opposed to solo – requires real consistency and precision. Across the aerobatic community, poorly flown barrel rolls hold the greatest risk to life for aerobatic pilots. Slowing the roll rate down in the second half of the roll can lead to unrecoverable nose-down attitudes and has, sadly, taken the lives of many aerobatic pilots around the world. The Red Arrows only ever roll left, *never* right, thereby avoiding errors of cognitive failure and potential mid-air collision. As Red 1, establishing the correct roll rate is fundamental to the manoeuvre's success. Roll too fast and the team simply can't hold on. Roll too slow and you can end up with a fateful nose-down trajectory. And finding the correct roll rate takes practice after practice. There is no shortcut to this aspect of training – the motor skill must be embedded in the leader's psyche. On a calm-wind day, with clear skies and a good horizon to keep you orientated, the barrel roll is a joy to lead; on a blustery, bumpy day with a poor horizon, this basic manoeuvre can sap every ounce of flying capacity from Red 1. If the conditions are too poor, the leader can decide at any time not to roll, and this kind of dynamic risk assessment is always in the forefront of Red 1's mind.

In leading the formation, Red 1 must also narrate everything they do to the rest of the team, voicing each input they put into their controls via the radio. In the air, all eight pilots in the team focus on the leader's aircraft; even those at the back or on the far side look through the aircraft next to them, in order to fix on references on Red 1's jet. At the same time the pilots must also follow the leader's voice-commands, which are delivered to a particular rhythm and intonation, a 'metronomic cadence' as the Red Arrows call it. In training, the pilots make inputs into their controls at an exact point or syllable of each command, with Red 1 and each pilot working to a well-rehearsed script and set of procedures.

A left turn, for example, is very simple to do, but when nine Hawks are flying only a few feet apart at 350mph, and the formation needs to turn gracefully as if it's a 'locked-together welded wing', each pilot must time their inputs into their controls with absolute perfection. The only way they can achieve that precise timing is by voice control, or 'flying by ear'. So when turning left, you might hear from the leader, 'Coming . . . left . . . *now*,' and Red 1 will deliver the words in a typical rhythmic tempo, perhaps with a slight rise in intonation or emphasis on the word 'now'. The word 'Coming' announces that they're about to do something, 'left' indicates which direction, and on the n of *'now'* the pilots need to make an input into their controls. Aircraft that are furthest away from the leader may need to time their inputs earlier, slightly before the n of 'now'. Red 1 must also give precise demands for when to put smoke on – such as 'Smoke on, go!' – as those smoke trails should look as neat and regimented as possible, with the team leader acting almost as a conductor, timing every move and colour to a well-rehearsed beat as the Red Arrows perform their own version of a symphony in the sky.

This narration of moves when flying is not something that military RAF pilots usually do, so leaders are usually keen to nail as soon as possible what can feel like an alien skill. During the first couple of weeks of being Red 1, I took every chance I could to practise, and I remember warming up my voice during early-morning showers, trying out such calls as 'Coming . . . left . . . now . . . and . . . pulling *up*', as I attempted to get the cadence and metre of the delivery just right. I'm sure my other half thought I was losing my marbles, from the outset. Thankfully, by the end of week one or two of training I felt fairly comfortable with the voice-commands, and they are of course central to the Red 1 role. Outside the specific displays, it can sometimes feel a little cumbersome leading a large formation – for example, when you're manoeuvring the team away from an airfield or lining them up with the runway. When flying something like a Tornado on your own, you would normally simply turn the jet right on its heels, pull the stick back and point it in the right direction. When leading a formation, however, you need to ensure that every aircraft is lined up correctly, and you have to narrate every roll and turn – it all takes that little bit more time and, in that sense, you can feel rather shackled to the other aircraft.

Once Red 1 has flown solo, perfecting the basic manoeuvres and the narration that goes with it, he or she will fly with new team pilots, usually Reds 2 and 3, who will then be joined by Reds 4 and 5 to form Enid. For the remaining team members, the process of getting used to a new Red 1 can be a little odd, having over the last year or two become entirely conditioned to the voice intonation of the previous leader, along with their subtleties in flying and general style of leadership. To feel comfortable with a new boss – someone with a different voice, who has yet to get to grips with everything, unlike the previous incumbent, who by year three of their tenure will have cracked the role – is quite a challenge, particularly for a team who are so bonded in what they do. It is something the whole team is aware of. As John Rands, former team pilot and Red 1 from 1994, put it: 'All leaders are different – you all fly slightly differently, speak differently, the coordination of voice with actions is different and the team learns to manoeuvre on the boss's call.'

BLENDING OLD AND NEW

Along with getting to know and flying with the team pilots, it's the responsibility of Red 1 to design and choreograph the coming year's show. The leader is aware that a display must include old favourites – those tried-and-tested crowd-pleasers – but also some new and upgraded manoeuvres, all of which must deliver the expected Red Arrows wow-factor. Displays must cater for the huge variety of spectators, from families at the beach seeing the Red Arrows for the first time, to aviation enthusiasts, some of whom really know our displays and look out for that bit of innovation every year – and as Red 1 you want to deliver that. All this is made doubly challenging for a new Red 1, who is attempting to follow on from the previous leader, whose show in their final year is likely to have been their best. After having had three years to finesse routines, the timing and links between sections will have worked well, making up a tight and geographically well-centred show that didn't see pilots needing to fly a gazillion miles away to make a turn and leaving a great big gap in the sky in front of spectators.

Weeks before starting as Red 1, I couldn't help but mull over the different shapes and manoeuvres that I knew. I remember being on holiday with my family and sketching ground-patterns out on tiny pieces of paper: drawing loops and rolls and linking things together, as I visualised how I might design the show. This part of the job can become rather obsessive – I have memories of being told off at the dinner table for sketching manoeuvres on a napkin with my children's crayons – but it is hugely rewarding when a concept in your brain starts to translate into a complete aerobatic display. A new boss will invariably bring his or her personal preferences and experiences to the choreography, and I had in mind a few shapes that I had enjoyed flying as a team pilot. At the same time, in the Red Arrows you are always mindful of what has gone before and which moves go down well with the crowd; the goal is to uphold the legacy and reputation of the Red Arrows while also injecting into the mix a bit of your own flavour.

As a team member I loved flying the Blackbird shape, named after one of my favourite planes, the iconic Cold War Blackbird SR-71 reconnaissance aircraft, and I knew the other pilots enjoyed training for it. It's a manoeuvre during which the whole team moves into one of the longest and sleekest formation shapes, in which Reds 1, 6 and 7 form the stem, leaving the rest to outline the elegance of the SR-71, so I was keen to include it in some way. I also wanted the jets to arrive at the shows in a different way. The signature entry for the Red Arrows had always been to arrive from behind the crowd in Battle formation – the classic V-shape formation – and then pull up, moving into a Diamond Nine formation as they climb. I wanted to try out the Wall – a manoeuvre that I remembered former Team Leader Jas Hawker mentioning he would have liked to do – in which the jets arrive spaced out really widely in Line Abreast, before moving into a short Diamond formation.

Introducing any revisions or new manoeuvres is always a collaborative process, one that involves discussion with the team and with the higher chain of command. The safety aspect must be thoroughly analysed, and senior officers must be reassured that every potential risk has been thought through. In doing this, Red 1 must design a training plan to incrementally break down the way they intend to train and then build up any new manoeuvre. Thereafter the team must work through the training plan, finessing, discussing and solving any issues as they go, until the move is performed safely and consistently and works well within the whole routine. In the earlier days of the Red Arrows, team leaders regularly introduced new manoeuvres and were required to get permission from the hierarchy before doing so, although the team would often try it out first in the air – very gradually and professionally, knowing what the capabilities of the aircraft and the pilots were – before presenting it for formal authorisation.

For the 2016 season I was keen to introduce the Wall and explained the concept to the team during training so that we could work out together how we would fly it. We went through eight weeks of attempting to fly it, plus constant discussion and revision, but we couldn't get the widths between the aircraft perfectly aligned. We were probably about three days away from abandoning it altogether when we decided to put tiny white markers on the back of each aircraft to help visually align each formation member, which ultimately solved the issue and saved the Wall from being discreetly shelved for another day. Overall it was a great relief because once we had cracked it, we loved flying it.

Many people on the ground cheered the team's new arrival shape and we received numerous favourable comments, although not everyone took to it. I distinctly remember being told by a Red Arrows enthusiast at the RIAT that it was a travesty that we had moved away from the 'usual' form of arrival and it represented the worst part of the show – many you will please, but not all. The Wall also made for a very apt arrival shape for the Red Arrows' first-ever tour of China, which I led in 2016, as it honoured the Great Wall of China. The tour there was a huge success for the Red Arrows, the RAF and the whole country. But for me personally, it was immensely satisfying to see the concept that I first scrawled on a tiny scrap of paper brought to life in the skies of the Zhuhai Air Show.

As it often takes the team several weeks to work out new

manoeuvres and shapes safely, Red 1 will usually bring in new concepts gradually over two or three years – and sometimes not at all in the first year, if the team has gone through any difficulties due to lack of aircraft available for training or exceptionally poor weather in the winter months, for instance. Another manoeuvre that I wanted to bring in, and am particularly proud of, is the Tornado shape, which I introduced in my second year as Red 1 in 2016 and adapted in 2017. I wanted to introduce something dynamic to the first half of the show, and I had seen the UAE aerobatics team Al Fursan perform something similar. The Tornado manoeuvre features the front seven aircraft flying in an Arrow position while Reds 8 and 9 perform very tight barrel rolls, so that they 'hug' the trailing smoke of the seven aircraft in front. For Reds 8 and 9 it's an incredibly difficult move, as each aircraft needs to be on either side of the tight circles they are flying. They have to employ a basic air-combat technique called 'lead and lag' – a method whereby you can 'catch' an opponent inside a geometric circle by leading at a certain angle – while ensuring that they don't rattle back or go too far forward away from the formation, or from each other. This is hard enough when they are performing tight little circles, but when the formation turns a corner, the circle then becomes egg-shaped and they can only achieve the rolls through visualisation. To achieve that kind of complexity, Reds 8 and 9 had to modify their techniques incrementally, involving extensive practice and repeated sorties.

In a similar way to the Wall, it took a long time for the team to crack Tornado, and when you're flying displays with some sixteen to twenty manoeuvres in total, Red 1 needs to carefully balance how much time is spent on each one. In trying to balance our training time, I cautioned Reds 8 and 9, Flt Lts Mike Bowden and Emmet Cox, during a debrief that we had two days left to smooth out any issues with the revised look to Tornado, which involved adding a 90-degree turn to the manoeuvre in 2017 – otherwise we'd have to scrub it. Faced with the stark reality of the situation, Mike and Coxie replied simply with, 'Right, boss, leave it with us.' They went off to an office, and I don't know what they said to each other, but on the very next sortie they rolled perfectly round the corner and we knew we had finally nailed Tornado. There aren't many RAF pilots who could do that, and it is a testament to the quality of the flying skills we have in the Red Arrows team. Tornado has become a real crowd favourite, looks beautiful in the sky, and the team still flies it to this day, albeit with a few revisions in 2020 (Reds 1 to 5 flying in Line Abreast rather than as an Arrow, with Reds 6 to 9 behind in a 'Tango' or T-shape). The Tornado makes for a fantastic image and features prominently in photographs that are shared by crowds on their social-media posts.

In designing the shows, I also wanted to create a good second-half manoeuvre, something akin to the Spaghetti Break, a spectacular move that sees the nine aircraft fan out from the top of a straight loop before landing, but which is only flown before arriving at some airports as it can't be flown over a display crowd. As a result, I decided to modify the Vertical Break, which involves aircraft looping upwards before breaking while pointing vertically down at the ground, producing a big fan of smoke as the jets separate outwards from one another. We had never been able to do that with all nine aircraft, so I put that concept to the team and asked for some creative solutions from them. They came back with a few ideas, one being the Hammerhead Break, which involves pulling up for a Quarter Clover (a twist in the loop to ninety degrees), then banking back, pointing at the ground, at which point we would split. It was the solution we needed and provided a really good transition from the first to the second half of the show.

Team leaders also encourage the team to come up with their own creative ideas, as synchro leader Jon Bond did in 2020 when he designed a way to adapt the Heart shape, a much-loved part of the Red Arrows displays performed by the synchro pair (with either Red 8 or 9 flying the spear that crosses through). Reds 6 and 7, however, require 6,000–6,500 feet of airspace to create the shape as they fly vertically upwards, meaning that the Heart can't be flown in displays with low cloud or airspace restrictions. To get round this, Bondy came up with the concept of angling the Heart shape differently, so that it could be flown in 2,000–2,500 feet of airspace. He designed it, put a safety case together for it, got the go-ahead to try it out and then flew half of the Heart on his own and tinkered with the technique. He then flew with Red 7 (Gregor Ogston) in the back seat and then, as a pair, they further perfected what is now the called the Rolling Heart shape (as it can be performed as part of the rolling display).

Manoeuvres and shapes are often specifically designed to tie in with special events, as was the case with my successor, Martin Pert, who designed a couple of manoeuvres to mark the centenary of the Royal Air Force in 2018. These included the Centenary Split manoeuvre, which is a revised version of the Palm Split; and the '100' shape. To create the '1', which is an impressive 5,000 feet tall, an aircraft must fly vertically from 100 feet above the ground, then two other aircraft pull huge loops to form the two zeros – all of which can easily look misshapen without extensive practice and assessment from the ground. Red 1 then leads in the remaining aircraft from the right, trailing smoke, to underscore the huge '100' in the sky. For the 2012 season, team leader Jim Turner revised the team's best-known formation shape, the Diamond Nine, in a bid to create something special for the Queen's Diamond Jubilee Fly-pasts of that year. Named the Deep Diamond, the shape was given a more three-dimensional feel, with aircraft stacked at different heights so that the Diamond shape could be seen from the ground as the team came in and out, rather than only when the jets flew overhead.

As leader, you're always on a sharp learning curve when it comes to designing shows and there's a lot you can learn from past displays. My predecessor, Jim Turner, was brilliant at keeping the first half of the show under constant G, which helps to keep the display really tight. Compared to Jim's last year as leader, the display in my first year was probably a little 'baggy', but by year three it was as centred to the crowd as I could make it. I knew I wanted to bring in the Infinity Break, an innovative and spectacular formation break that was designed by my former boss,

Jas Hawker, and had always proved an awesome way to finish the show. However, I had parked that for my third year, as it's a challenging move for Red 1, who rolls around the smoke of the team to create the shape of the infinity symbol before the formation splits. The Infinity Break is a little like patting your head and rubbing your stomach: as the leader, you must ensure that you line up the team on the correct heading, at the exact speed and at the right angle down, then safely reverse back, fly two perfect barrel rolls, get the radio transmission for the break out at the right time for you to come through the middle of Red 2 and 3's smoke, while pulling straight up at 5G. And, of course, you must be ready to execute an escape manoeuvre, should anyone in front of you unexpectedly have engine failure. Again, it takes practice.

INHERITED WISDOM

As Red 1, it's vital to draw on the experience of previous leaders, because only they have experienced what you are going through and know well the challenges that you face on a daily basis. One of the most informative days of my training as Red 1 was the annual ex-team leaders' conference, which was attended by nine or ten former team leaders, from 1972 to the present day, at RAF Scampton. Some of the former Red 1s who pass the medical fly in the back seat of the Hawks, and Group Captain Ian Dick, who was leader between 1972 and 1974, was going to join me in mine. While he flew the Gnat during his time with the Red Arrows, the techniques of flying displays then and now are much the same, so there's always a lot that team leaders can discuss and compare.

With Ian in the back seat, I flew a full show in the Hawk, after which he gave me a few pointers. 'The way you entered the barrel roll,' he said, 'was just how Ray Hanna taught me to do it' – and to hear that one of the founding Red 1s might approve of the way I flew was very reassuring. The comment 'Nice roll rate, good finish with a smooth let-out' was also music to my ears. To have a senior officer say something looks good or safe, or for Reds 8 or 9 to let you know if something is missing from your performance, is one thing, but to have such affirmation from a former team leader is quite another and it was incredibly gratifying. If I hadn't quite nailed something, Ian wasn't shy about telling me: 'I think you're looping a bit high – you can probably chop five hundred feet off your Blackbird to Fred twist', and it was equally useful to hear this.

During the ex-team leaders' day, ex-Red 1s also join the existing team on the ground, sitting in on briefings. After that, the team leaders (both past and present) shut the doors and talk amongst themselves about a host of issues, ranging from the kind of problems they faced as leader, to how they dealt with different personalities and issues within their teams. They might also give you a bit of feedback about your own team, having spotted that perhaps Red X or Y isn't communicating enough in a debrief and they would have wanted to hear more from him or her, or that a certain pair are communicating really well with each other; along with various other comments about the general dynamic of the team, which is so critical to the performance of the Red Arrows.

Covering everything and anything, outside the actual nuts and bolts of flying, former team leaders might then discuss their experiences of dealing with superiors, which has always been a consuming aspect of being Red 1. The leader role is multifaceted, requiring not only technical expertise in the air, but also diplomacy on many levels. And the reputation of the Red Arrows is so important to the RAF that the reach into the squadron comes from the very top – in some ways more than in many other squadrons. Incidents, or reviews of general defence funding, can lead to a more forensic analysis of operations and costs, and for some leaders the constant back and forth with senior officers can be the most demanding aspect of the job, particularly in periods when the future of the team has felt precarious – and especially before 2012, when Red 1 had the dual role of team boss and Officer Commanding.

John Rands, Red 1 from 1994 to 1996, well remembers the pressures of the job: 'Every year the military has a pot of money, and the Reds stand out because it's not a front-line unit, so it's always under the microscope, and how do you quantify the worth of an aerobatic team? With the RAF not needing to recruit as much, we were conscious it could end at any point, but it never happened and we were ultimately left alone. The flying was the easy bit.'

While the general public and aviation enthusiasts have always been constant in their affection for the Red Arrows, the team is aware of its perceived image within the Royal Air Force. In the first few years of the team there were those who deemed the Red Arrows 'flashy', its pilots aloof and arrogant – so much so that at some events pilots avoided wearing their red flying suits if they could. Although the Red Arrows team undoubtedly selects pilots based on merit, and in that sense it is an 'elite' unit, the dynamic of the team simply doesn't allow for big egos and – as one former leader found – just a day spent with the Reds was enough to dispel any prior misconceptions that he'd had about the team. There is now much better internal messaging within the RAF, but that negative portrayal does occasionally raise its head, perhaps among those who underestimate the skills and hard work that go into flying with the Red Arrows as well as the relentless tempo of the team, not to mention the role the Reds play in representing the UK and the wider defence tasks of the Royal Air Force.

As the public face of the RAF, the Red Arrows are frequently under the media spotlight, which often shines most brightly on the team leader. Switching from being Red 1 in the air to an interviewee on-camera – sometimes almost as soon as you have stepped out of the cockpit – can be difficult, and it's always interesting to hear from other former Red 1s how they coped with that rather strange mix of pressures. On my first global tour in China some of the interviews that we did on their state-owned breakfast TV reached around half a billion people, and so the advice I received – 'Don't think about the enormity of the audience, just talk to the person behind the camera'– proved invaluable.

In many ways, compared with most flying roles in the RAF, this area of the job is a great example of how broad Red 1's remit can be. Being offered the chance to speak directly to a diverse range of audiences is in itself a strategic capability: your responsibility is to message on behalf of the RAF or, depending on where in the world you are, the UK. After a demanding sortie it can be a challenge, but you soon stop worrying about the size of the audience and think instead about *who* your audience is, and that you must communicate authentically, while making sure you remain on-message. Indeed, the role involves appearances across a whole range of social-media platforms and, for a fighter pilot who normally has little contact with the media, the team leader has to adapt quickly to rise to this new challenge.

THE BUCK STOPS WITH YOU

Managing such a close-knit team of pilots also requires a heavy dose of emotional awareness and competency, utilising many of the skills that I learned as an officer and flying instructor. Before the team heads off to fly, we ask at the briefing, 'Is everyone fit to fly?' It's a verbal contract that every pilot must answer 'Yes' before we proceed any further with the technical aspects of the subsequent sortie. However, life is never entirely clear-cut: pilots might be fit to fly, in that physically they are up to it, but there may be more subtle elements at play that might indicate they are not at their best. You can pick up on how they might be feeling that day just by looking into the eyes of the pilots, or even by watching their body language or posture, or by conversations that you overhear. If some of those eyes look a bit darker, simply as a result of tiredness, as Red 1 you might consider how far you want to push pilots that day. You might sense that a particular pilot is not as buoyant as usual or there is some anxiety there; conversely, someone may seem a tad over-confident or bullish. And all of this you try to explore in the briefing room.

Just as the ability to accept criticism is vital for team pilots, so it is also essential for Red 1 – everyone has to be honest in the debrief, and the leader needs to own up and correct any mistakes, particularly in their first year when still getting to grips with leading the formation. Good communication between engineers, support staff and pilots is also critical to the smooth running of the Red Arrows team, and the team leader needs to be aware of any issues that might affect the performance of the team as a whole.

Fundamental also to the team's success is an open and honest relationship between Red 1 and the Officer Commanding (OC, who flies as Red 11). From the team's inception right up until 2012, Red 1 had the dual role of team leader and Officer Commanding. Supporting the leader was a Wing Commander (Red 11), who dealt with many of the organisational issues required to keep the team going and who ultimately dealt with numerous formalities of squadron life that Red 1 didn't have time for while flying three times a day. However, following a significant overhaul of military airworthiness regulations in 2010, the decision was taken to align the Red Arrows team with the more traditional command structure of most other RAF squadrons, where a Wing Commander assumes responsibility and accountability for squadron output. While the team leader has to focus in real detail on specific issues, the Wing Commander – who is the Officer Commanding and is more removed from the pressures of constantly leading the formation – can provide a broader perspective on matters, or simply act as a confidant and sounding board for Red 1.

During my time as Red 1 I found the support and mentorship of my OC, Wing Commander Martin Higgins, invaluable in many instances and was very grateful for it. Now, in my role as OC, I realise that my job (aside from leading the entirety of the squadron) is to protect Red 1 and ensure that all the organisational stressors that are often in play are not placed solely on his or her shoulders, thereby enabling the leader to focus entirely on the challenges of their flying role. When the team is at home base during the training work-up, shielding Red 1 from wider issues is relatively easy to achieve. When the Reds are deployed for the majority of the display season, Red 1 is dynamically making risk-based and strategic decisions, both in and out of the cockpit. Although the OC will travel with the team to many events and will provide a metaphorical umbrella at the more high-profile engagements, the pressure to perform at the expected levels associated with the Red Arrows team have not changed for Red 1 in the modern era. However, with the OC bearing accountability for the overall output delivered by the team, the leader always has the freedom to pass along any organisational issues to the OC, in order to reduce unwanted distraction and remain in the right psychological mindset, especially when airborne.

In leading the formation, Red 1 must also try to present a calm, assured presence to the rest of the team, which can be another challenging aspect of the job. There were times when I drove to work, perhaps at the end of a gruelling week, and I'd think the last thing I wanted to do was strap on the jet three times. I would know that the day ahead was likely to be tough, but as I walked down the corridor to the briefing I'd try not to show too much of what I was dealing with personally. I would 'put my face on', as it were. When you have a group of pilots who trust you implicitly, both on the ground and in the air, they require a steady hand, and consistency and composure are a must.

In an attempt to acquire the right mindset before briefings, I'd try to create a little fifteen-minute bubble beforehand: so no looking at emails, and only emergency calls on my phone. Occasionally I'd have to come direct from another meeting, or break my own rule and look at an email two minutes before a briefing, which then might trigger feelings of irritation or simply distract me from what I was about to do. But even if you're seething inside, when you go into the briefing room as Red 1, you somehow need to park whatever you're feeling and focus entirely on the flying objectives and potential hazards of the next sortie. If you let those negative feelings take over, you might react in the wrong way to something in the briefing room, simply because you were annoyed by an email that you read a few minutes before. If you respond unpredictably to the occasional distraction, you've lost consistency and you're not quite the steady ship the pilots all thought they were safely tethered to, and the team's confidence in you may begin to falter.

While the single point of failure rests with Red 1, and the show does not go on without the leader, he or she must never fall into the trap of believing that the role is bigger than the team. After three years as leader your tenure is over, and the reputation of the team always comes before you as an individual. You don't drift off into total obscurity when you leave, but the new leader does take the spotlight and must now drive the team forward to new and exciting manoeuvres and places in the world.

Each of your three years as Red 1 comes with immense highs and lows, with the well-worn calendar providing a wealth of mile-stones and stand-out moments. Securing the PDA at the end of training is a huge moment for any team leader – I remember the euphoria I felt on earning my first PDA, whereas my team leader at the time, Jas Hawker, always seemed very cool and collected about the whole affair. Euphoric as you are, you are also relieved that you've made it through winter training, but the curse of the job is that you're already thinking about the next event and the busy display season ahead, plus all the problem-solving required by you to make it work. The first public show in front of a crowd is an immense day, as is your first international transit, when you have overcome the logistical nightmare of getting all our Hawks and the entire team overseas in one piece. Your first fly-past over Buckingham Palace, or any of the major cities around the UK, is a huge marker in the year – and who could forget the many special events and anniversaries that we take part in around the country.

You are reminded, time after time, that being Red 1 is one of the best flying jobs in the world (and without doubt it is), but as the team leader – 'the boss' – you know that ultimately the buck stops with you, and that the consequences of something not going right are huge. It's something you never stop thinking about and, as leader, my entire time was a battle between striving for the joy of success versus the constant, nagging fear of failure. Home life often has to fit in around the relentless Red Arrows schedule, which is particularly tough if you have young children at home, as their needs don't necessarily tally with early-morning training sorties or display season; and – as with all military families – it's often your partner who has to hold the fort at home, something I'm forever thankful to my partner for. During your tenure as Red 1, the Red Arrows team becomes your other family, and the relationship between all nine pilots is intense. Holding all of that together, while maintaining your reputation and constancy throughout, is draining.

Life on front-line squadrons of course comes with its own pressures and risks – not only are you acutely aware of all the sophisticated (and rudimentary) missiles, guns and rockets that an enemy can shoot at you, but the lives of many soldiers, sailors and air personnel can depend on your actions. But three years of being Red 1 does take its toll, and at the end of my tour I was in desperate need of a break, away from that knife-edge existence where the wrong decision or the slightest nudge on the control stick could lead to disaster. Adrenaline gets you through it, but the exhaustion does creep up on you, as Dickie Duckett found, on leaving the team at the end of a busy and very hot display season in 1976: 'After forty-five displays over July and August, flying to the limits, getting the whole team from A to B, getting down on a runway through cloud, the pressure mounts and you don't actually realise how tired you are.'

Nonetheless, while you're in the job there are plenty of upsides, not least that you are fulfilling a childhood dream in one of the best roles that a pilot could ever hope to have. Troublesome issues, or disagreements between pilots on the ground, tend not to linger after sorties – and there's nothing like punching through the cloud at 5,000 feet to give you a bit of perspective on life. When you're flying three times a day or engaged in a busy schedule of displays, issues that might have blown up tend to be resolved or to have fizzled out by the evening. The Reds are also performance-based, and we get immediate daily satisfaction from a job well done – unlike many roles in the RAF or in civilian life, where you might be working on long-term projects and might not see the fruits of your labour for weeks or even years. And, of course, you get to spend a large chunk of your day flying, which for a Squadron

Leader is unique to the Red Arrows, as former Red 1 John Rands found: 'There is no other unit in which the boss flies that much. Elsewhere, the lion's share of flying goes to the young guys, and the boss (normally in his or her late thirties) would have meetings, staff work and paperwork to focus on. Leading the Reds, from very early morning to late in the day, you are briefing, flying, debriefing, three times a day – it's a definite perk of the job.'

Despite the many challenges of being Red 1, the pure joy of flying a full display in sky-blue conditions represented the best days in the job, when all the mounting pressures simply evaporated. There is something incredibly special when you and the team are perfectly in sync, with the pilots focused on the metronomic, almost meditative cadence of your radio calls, almost as if you've stepped off the world for a short time. Many times, cresting the top of a loop surrounded by red jets, I'd allow myself to take in that second and think, 'There is no single place on the planet I'd rather be right now than leading this team.'

7

The Check-in

THE RED ARROWS showcase the very best of precision flying at their public shows, performing an array of breathtaking manoeuvres, much to the delight of spectators below. It is critical, though, for the team to arrive on time, to the absolute minute and second – whether it's for 50,000 spectators assembled on a breezy beach in Wales or for a fly-past over central London, the clock rules everything.

The process that ensures that nine Hawks arrive exactly when they should, with our signature red, white and blue smoke trailing behind, begins with the Red Arrows' 'check-in'. Renowned throughout the RAF, the Reds' check-in time is set in briefings and builds to the moment when the team pilots are in their aircraft, ready to receive the 'Red Check' call from their team leader. Every flight, both in training and during the display season, begins with the Red Check-in: the point at which every pilot in the formation confirms that he or she is primed and ready to go. On a display day, the time of check-ins must correlate with a complex minute-by-minute schedule known as the WHAM (which stands for What Happens According to Mange – 'Mange' being an affectionate term we have for the Team Manager, who traditionally writes the schedule). This schedule dictates what happens and when, and everything from when personnel should arrive at base to flight arrival times.

Red 1's briefing of pilots will start to the absolute second, during which he or she will run through, visualise and discuss with the pilots the display sequences, weather forecasts, potential hazards and emergency procedures. The whole process is run with minimalist efficiency, is kept to the point and without any chit-chat or time-wasting; it is held either in the briefing room at RAF Scampton or, if the team is out on the road, anywhere from inside a moving minibus to beside a Hawk jet. I've even briefed the team under the wing of a Boeing 747.

At the end of the briefing Red 1 will give the check-in time, to the second, by which point pilots should be in their aircraft, ready to receive the leader's call to start their engines. Pilots are now time-stamped and are given around a twelve- to fifteen-minute window to run through a prescribed and well-rehearsed list of procedures that ensures team members maintain their essential focus. Within that window there is a lot to fit in, all of which must be done safely, and one of the biggest crimes for any Red Arrows pilot is to be late for a check-in.

As soon as they leave the brief, pilots must get their kit on – usually after a quick toilet stop. Each pilot will already be in their red display flying suit, but they now must put on their lifejacket (which contains various survival equipment, including a first-aid kit and mini-flares), gloves, as well as anti-G trousers and an individually fitted helmet, complete with an oxygen mask and internal headset for communication. For me, putting on the kit was an important part of getting into the right frame of mind for the sortie – in particular the anti-G trousers, as they are such a close fit to the body. Gearing up always prompted a subtle mindset shift, so that I was mentally and physically ready to proceed.

Once all the kit is on, pilots walk to the engineering desk, where there is a large folder that is familiar to all RAF pilots, known as the F700. Pilots must check that their aircraft – each Red Arrow is assigned his or her own Hawk, equipped with everything it needs – is full of fuel, coloured dye and oxygen, amongst other essentials. If all is present and correct, the pilot 'signs the aircraft out' – a legal requirement that indicates the pilot has fully understood that the aircraft is serviceable and ready for what they are about to. In doing so, the responsibility for the aircraft passes from the engineer to the pilot.

STRAPPING ON

Walking to my Hawk, I often felt the adrenaline start to build. What I love about the Red Arrows is that – and this is very different to front-line flying – wherever you are in the world, all the aircraft will be parked perfectly in formation, expertly aligned to each other. That special sight always made me feel that the Red Arrows are a collective. It is never about what I'm about to conquer in the jet, it's always about what we will achieve, as a team.

As team members approach their Hawks, their Circus engineer (one being assigned to each aircraft) is there waiting for them. They exchange pleasantries and maybe a bit of light banter – a very different experience from the American aerobatic teams, who salute their engineering officers in a far more formal see-off. In the Reds, Circus engineers and pilots generally have a professional but easy-going relationship, and those quick, light-hearted exchanges can help to take the edge off any nerves.

Pilots will then do a walk around the aircraft, making specific checks and ensuring that the physical components of the jet, including the control surfaces and hydraulic pressures, are all in perfect order. At this point the jet is completely dormant – all is quiet and it has no life in it whatsoever. When you're a pilot on the front line, as you walk towards your fighter jet there might be a unit already giving the aircraft electrical power and there is more of a humming noise or background throb – the kind of noises you might hear as you climb up the steps of a commercial airliner. With the Hawks, you don't have any of that; they are entirely silent and inert, simply smelling of the fuel that has just been pumped in.

Walking around the Hawk, looking at its airframe, I'd often think about all that potential energy: how something so apparently lifeless was about to do something so phenomenal. At other times, seeing the Hawks parked up in the hangar, I'd marvel at where those airframes had flown over their four decades with the Red Arrows. These jets had been all around the world, from minus-twenty degrees close to the Arctic Circle to the sweltering heat of the Arabian Desert, the jungles of Vietnam and across China and Australia – and they had weathered it all and travelled so far; and there they sit, lifeless and yet ready and waiting for you to unleash them on the world again.

Red Arrow pilots tend to have a real connection with their aircraft: their name appears on the side of their plane, and they know every inch of it, more so than any other aircraft they have flown before in the RAF, because they take to the air so often in it – three times a day, five days a week, and even more often during the display season. It's all about raw flying when you're in the Hawk or the Gnat, and the intimate relationship you've built up with your aircraft – something that is, at the end of the day, merely a lump of metal and cables – is far closer than with any other type of aircraft.

Once all the external checks have been done, pilots climb into the cockpit of the Hawk, which is much smaller than that of modern fighters. Next you put on your helmet, before connecting yourself to the umbilical cords that join you to the aircraft: the oxygen, anti-G protection and the microphone from your mask. You then fix the leg restraints, which keep your legs secure in case you should need to eject; and then put the main straps over your legs and shoulders, buckling them down in a process that feels more akin to strapping the aircraft on to you, rather than the other way around, as if you're wearing a five-and-a-half-ton backpack. You are now entirely connected to that aircraft – I'm pretty sure you can't be more physically attached to something mechanical – and you can't help but think, 'Right, it's me and you now.'

Once you're fully strapped in, you turn the internal batteries on and the aircraft starts to come to life. The emergency warning panel gives a few welcome bleeps to let you know it's working, then you check the throttle is functioning as it should and, with the engineer's ears close to the airframe, you check that the igniters in the engine are working correctly. You then do your 'left to right' pre-start checks, moving through all the instruments in front, from your left to your right side, switching on the radios, navigation aids and everything else that sets up the jet for engine start. You are now ready for the moment you've been building up to since the end of the briefing: the Reds' check-in. You select the standby radio, counting down the seconds with your watch, and with your thumb over the transmit button. Red 1 will call 'Red Check' and each pilot replies with his or her respective number in consecutive order – '2, 3, 4, 5 . . . ' – which confirms that every pilot is there and that, crucially, their radios are working.

Red 1 can often pick up on the mood of the team just by that check-in. At the end of a difficult day, perhaps when you're flying multiple shows, the delivery of those numbers might be given in a perfunctory manner, as if it's simply another day at the office, yet the pilots are still fully aware of the immense responsibility of the job. However, on days when you're off to your first show and you know there will be 400,000 people on a seafront waiting for you, those numbers might be shouted out with real gusto, indicating to Red 1 that the team is well and truly up for it. Once everyone has checked in, Red 1 will call for 'engine start' and the whole team will go through their engine-start procedures. The Hawk feels like a living, breathing thing, the needles in the cockpit are moving and you can sense all that raw potential energy about to be released. The anticipation mounts: 'Here we go!'

Once everyone has their engines lit, Red 1 will taxi-out first and the rest of the pilots will follow in perfect formation. That's the plan anyway, but sometimes keeping in that shape can be tricky, with the Red Arrows occasionally losing a little of their usual airborne grace on the ground. After all, aircraft are designed for flight and can often look a bit cumbersome travelling over terra firma. The Red Arrows must negotiate a variety of airfields, and not all taxiways are perfectly flat; some have ramps and slopes that can catch pilots out. As a result, a couple of Hawks might fall off the back and, when pilots unwittingly deviate from their taxi reference, a call such as '5 long' will be made, to ensure that the rest of the formation is aware of the others' movements or unplanned deviations. When things aren't looking as smart as possible, the 'long' call may rattle down the formation, and the further back it's called, the grumpier the tone may become. If the call makes it all the way back to Red 9, who is being pushed backwards by the planes in front, then the call '9 uses power' can convey a sense of scorn to those in front, as if to say, 'Come on, guys, we're meant to be the Red Arrows, so sort yourselves out.' As with all aspects of the formation, it's Red 1 who sets the correct speed and is often responsible for any knock-on struggles at the back. If the taxi route looks very tricky, with minor climbs and tight turns, Red 1 always has the option of calling for double or (rarely) quadruple spacing between the aircraft, which helps the team strike the correct balance between presentation and safety.

TAKING TO THE SKIES

Once Red 1 has briefed the kind of take-off the team is about to do – whether a display take-off, a transit take-off or a stream take-off – the Hawks line up on the runway. I loved this moment when I was a team pilot. As Red 1, however, this was when my mind raced, thinking about what lay ahead. There is no chat amongst pilots at this stage; other than air-traffic control, the radios are quiet, and it's all about focusing on the job. Reds 1, 2 and 3 will start rolling down the runway; four seconds later Red 6, leading Reds 4 and 5, will follow suit, with Red 7 leading Reds 8 and 9 eight seconds later.

Leading a formation during take-off demands the highest levels of focus and awareness, and the smoothest touch on the controls. Easing the aircraft at relatively slow speed into the air must be done with a positive, yet perfectly progressive control input. This means smoothly and gradually easing the control column backwards to make sure the aircraft leaves the ground properly. Making any change to the control column's position is called a 'control input'. Once airborne, Reds 4 and 5 will meet the boss at the front with 2 and 3, and Red 6 will lead the back section. You will fly in those two sections – Red 1 leading Enid, and Red 6 leading the rear section and always in visual range within half a mile of Enid – as it's a manoeuvrable flying formation that gives you the balance of effective lookout for other aircraft while maintaining formation integrity. You will then join together as a nine-ship formation about two to three minutes before a display or fly-past.

Transits to displays in the UK are generally done at low level, at anything from 500 to 2,000 feet above the ground, and when you're travelling at 350mph, getting from A to B is quick. For instance, it takes approximately twenty-five minutes from RAF Scampton in Lincolnshire to RAF Valley in north Wales. To transit to a display site, perform a twenty- to twenty-two-minute display and fly back on a single tank of fuel, the maximum range between take-off and the landing airfield is seventy nautical miles – that's just over eighty regular miles. If we are on a transit only, with no display, we can comfortably route at low level for much further distances: for instance, from the south of Scotland to central England or south Wales. On longer transits we will fly at a higher level of more than 20,000 feet, where the jet is very fuel-efficient – at 40,000 feet the Hawk burns only one kilo of fuel for every nautical mile (or 1.15 regular miles) travelled.

On a typical low-level transit the world starts to go by a little faster, but pilots are focused on the formation, listening out for air-traffic control or scanning for any other aircraft that might be a hazard to us. The leader, who can see Reds 2 and 3 in the mirrors and a little bit of Reds 4 and 5, will be making quick decisions on the route, second-watching the entire time, checking the speed they need to fly and any other corrections that have to be made due to that day's air traffic and weather.

In the cockpit your left hand falls naturally on the throttle, with the control column or stick in the right hand. Although pilots all have different techniques, many of them rest their right elbow on their thigh, tucked in with the arm locked next to the ribcage. This positioning enables them to hold the stick firmly, but with a light touch. Using this method, they can make really small inputs: the Hawk's controls are so incredibly sensitive that you can roll the aircraft just by pushing the stick to the left by a fraction of a centi-metre. Pushing the throttle forward and back controls engine thrust, and the pilot's feet rest on rudder pedals that move the fin at the rear, allowing for lateral control, either for keeping things straight or while executing a turn. Using your feet when flying is the norm for helicopter pilots, but fast-jet pilots generally never need to use a rudder to turn a jet; instead they will bank their aircraft on its side using the stick, then pull back and round the corner. In the Red Arrows, however, our movements must be so precise and accurate that we constantly control the rudder in order to drive the aircraft while keeping the wings level. When you're Red 9, at the back of the formation looking through three other aircraft at Red 1, the last thing you need to see is the other pilots' wing-tips twitching up and down.

Navigation equipment in the Hawk T1 is pretty minimal, being reliant on a 1980s-style GPS satnav bolted on to the instrument panel, which shows coastlines and airfields and gives you a bearing to steer to, depending on what you are aiming for. As the 1975 team leader Dickie Duckett explains, in the 1960s and 1970s navigation aids were even more rudimentary, involving 'getting a map out, drawing a line on it, measuring headings, using the stopwatch and navigating visually as you crossed the country'. But these low-tech methods still did the job – the team always arriving as scheduled, with an acceptable window of plus or minus five seconds. These days our GPS provides a primary navigation capability, but we still make full use of visual referencing to make sure we're in the right location. Our maps are now printed with the route rather than hand-drawn and, in the event that the GPS fails, our trusty stop-watch is always running in the background as backup.

For every transit sortie – whether it's a UK low-level quick hop or a two-hour international route in Class A airspace with commercial traffic – an allocated member of the team is given the job of planning it or, in Red Arrows-speak, becoming the 'Nav'. The responsibilities of the Nav can be shared by anyone in the team and, to keep things fair, the role is handed out evenly among the pilots over the course of the season. Depending on the complexity of the route, the sortie plan may take two to four hours to complete correctly. Once the plan is complete, the Nav will ensure that all pilots have the necessary maps and will call all the air-traffic control centres and ground-handling agencies that he or she intends to use, to give them a heads-up. This level of coordination ensures that the sortie goes as smoothly as possible and is crucial, because international airports are not generally used to handling ten or eleven fast jets in close formation. Thirty minutes before the main briefing for the sortie, the Nav will brief Red 1 separately on the entirety of the plan. This is done mainly to double-check that every-thing is understood, and Red 1 will already have received the maps ahead of this and will have had time to mentally visualise the sortie, especially if there are complications, such as bad weather.

In many ways, overcoming operational challenges such as poor weather conditions can make this process even more satisfying. As an example, in 2016 we were tasked with displaying at the Wales Air Show at Swansea seafront, but as we sat in Cardiff airport, all we could see were low clouds and rain. Red 4, Flt Lt Mike Bowden (the Nav for this sortie), was poring over all the streams of weather data, to work out how we could avoid letting down the audience with a no-show – and then he spotted a gap. In the satellite imagery he saw that a small hole of clear weather had developed west of Swansea over the Pembrey area, and between us we worked out a plan. With the help of the brilliant Cardiff air-traffic control, we decided to get airborne in three separate formations, straight into the cloud, and to use timing and transit heights to reduce the risk of collision. Reds 1, 6 and 7 were the section leaders, with two jets each on their wings. We all made it to the clear weather safely, caught sight of each other and then joined up as a complete formation. With the weather a little better in Swansea, we safely ducked under the cloud and made it to the display site. Sadly, on this occasion the weather wasn't good enough for a complete display, but we managed to fly a flat show and departed as soon as we approached our minimum fuel level. We climbed back up into the cloud, split the formation down into three sections again and made our way back to Cardiff. The weather there was still awful, so we flew instrument approaches, just as a commercial airliner would do, until we could see the runway. All three formations made it down and we taxied back home.

Over a welcome cup of tea and a biscuit, I remember feeling very proud of the team's achievement that day. We had used every ounce of our combined aviation experience to ensure that we safely delivered a show for the crowds on Swansea seafront. The Nav had built a good plan for the rendezvous; we had flown on instruments in thick cloud on take-off (still in formation) and when landing back at Cardiff; and, what's more, we had conducted precision display flying – I realised that we'd had to call on almost every demanding aspect of aviation in that crucial one-hour and fifteen-minute time period.

These days, as the Wing Commander, I will often join the formation in transit sorties as Red 11, and I have the benefit of seeing the full formation ahead of me: a synchronised mass of aircraft, seamlessly in concert with one another, with their wing-tip vortices sometimes trailing thin wisps of white behind them. If the weather is poor, there's no time for anything other than pure focus. Windy or gusty conditions can lead to what the Red Arrows refer to as 'the bumps', causing our aircraft to move around by two to three feet and meaning that pilots have to work extra-hard to keep themselves in position. The team specifically trains to deal with the bumps, but it's one of the most hated parts of the training process, as it can feel as if you're pushing the jet through treacle.

SHOWTIME

With Red 1 clock-watching the whole time, the team will near the display site and begin the 'run-in', which is usually met with an intense rise in adrenaline as pilots start to line up for the run-in track to the show. For new pilots, heading towards the first public display of the season is an incredibly tense period, as former Red 3 Kirsty Murphy remembers, when she headed to the team's initial show at RAF Brize Norton: 'Just before the "Smoke on, go!" I remember my heart was pounding. I really had to calm myself. When you fly in formation you can easily tense up, and then you tend to grip the stick tightly and your knuckles whiten. As instructors, we always used to say, "Wiggle your fingers" to relax, as you should never hold the stick too tightly.'

At this point the team will move into their arrival shape: the trademark Battle or Wall formation, with the jets moving so that they're twelve to fifteen feet away from each other – so close that pilots can read the writing on their partners' helmets and even their facial expressions. The run-in usually begins four minutes and around twenty miles from the site; and the team almost always arrives 'crowd-rear' – shooting across the heads of the spectators from behind – as we've found that this gives the best impact and wow-factor.

Red 1 is still computing every second, making sure they are all on the right heading, at the correct speed and have briefed the right details to the team. The leader must decide and communicate the type of display that the team will be doing, which is always dictated by the weather and cloud cover. In the intemperate and unpredictable climes of the UK, all sorts of weather can roll in from the Atlantic Ocean: we might see thunderstorms, bright sunshine and dark skies again during the course of a thirty-minute display. To factor in the vagaries of the British weather, the team flies three different types of display – full, rolling and flat – and is one of the few aerobatic teams in the world that can switch between different types of display during a show. A flat display, consisting of fly-pasts and steep turns, will be flown if the cloud base is as low as 1,000 feet. If the cloud base is higher (clear up to 2,500 feet), the team will perform a rolling display, with rolls and wing-over manoeuvres. For a full display the skies must be cloud-free, or at least clear to above 5,500 feet. These conditions enable the team to fly full loops without the danger of entering cloud at the top of a loop.

As the formation approaches the site – now with perhaps 120 seconds to go – Red 10, acting as display commentator on the ground, will confirm pressure settings and wind measurements at the display site and, after a quick bit of mental maths, will inform Red 1 which kind of display to go for. The leader will ensure the team has the right pressure setting on their altimeters, and then Red 1 will talk to Red 6, who will tell the leader how he will 'wind' the display, meaning which subtle adjustments will be made to ensure the display stays perfectly centred for the crowd. If Red 1 thinks they are clear to go for a full display, he or she will radio-call to the team, 'We are starting full.' On hearing this, if the transit has been problem-free and the run-in straightforward, it has become traditional for Red 1 to give the microphone to new pilot Red 2, who then has about five seconds to gee-up the team or deliver an apt one-liner that will, hopefully, make all nine pilots laugh. The team knows each other so well that usually Red 2 knows exactly what to say: perhaps an in-joke or a bit of friendly banter at the expense of one of the other pilots. In these moments I've seen jets aligned in Battle formation bobbing up and down, their respective pilots shaking with laughter. It's only done when the weather and other conditions permit, but it acts as the perfect ice-breaker – a quick and welcome antidote to the mounting adrenaline, before the team heads into a public display.

If the weather ahead looks variable, Red 1 may delay his call and may radio to the team at the beginning of the run-in, 'This will be a late decision.' This gives the leader the opportunity to further review the weather conditions, although he or she *must* make a call within ten seconds of their first manoeuvre, whether that's pulling up for a loop for a full display or turning left to go flat. The next thing to do is to get the smoke on and prime the tanks, so that they're ready to turn on the colour. When Red 1 can see the crowd, either on the coastline or at the airfield, he or she will give the final confirmation for the display, for instance, 'Starting full.' Now the team knows exactly which show they're going into. At the call of 'Smoke on, go!', the iconic red, white and blue colours will start streaming through the sky, the team now being in full presentation mode, to the delight of the spectators below.

While a show may start with one of three types of display, the team can (and often does) switch dynamically between those three displays mid-performance. Red 1 will have choreographed the show so that he or she knows at exactly which window of opportunity they can transition between the three types, and the team is trained to move seamlessly and safely from one manoeuvre to another. By the end of training, the script for each display and sequence of manoeuvres will be firmly embedded in the minds of each pilot, so that they know exactly what they're doing and can reduce the likelihood of any kind of cognitive failure or miscommunication during a display.

A display might begin as full show, with a big roll, but if there's cloud at one end of the display area, rather than following with a roll, Red 1 will radio, 'Going flat' to the team. Once that is acknowledged by Red 2 and the rest of the pilots, they will move into a flat turn. From there, they know which set of manoeuvres follows that flat turn. Within the same display they may be told to go full again or to move into a rolling portion of the show, before a full finish. In-the-moment adjustments to a display are challenging, but it's one of the many reasons why we drill our manoeuvres and transitions as much as we do, so that even a stitched-together display – with full, flat and rolling parts – can appear totally seamless. To the crowd below, it will probably look exactly like a normal

Red Arrows performance, but to the pilots it may well have represented the hardest twenty to twenty-three minutes of their lives.

A GAME OF TWO HALVES

The first half of a Red Arrows display consists of synchronised formation aerobatics, followed by a more dynamic second half, during which the synchro pair, Reds 6 and 7, perform their opposition passes. The team can form many different formation shapes, which are usually based around five reference positions – Battle, Diamond, Arrow, Line Astern and Line Abreast – although other formations, such as Leader's Benefit, may consist of a combination of those reference points. Larger formations will hold shapes in loops, rolls or bends, with those on the edges of the formation (in particular Reds 4 and 5) flying a longer path around, in loops and bends, to maintain the overall shape.

If the team is able to fly a full display, it will always start with a vertical manoeuvre (what's called the 'arrival loop'), as it makes an impressive start. The first loop of any full display is really key and often sets the tone for the whole show. As Red 1 calls, 'And pulling up' – which might also be relayed to the crowd on the ground commentary – the leader wants to see the team absolutely locked-on and not bobbling about on the wing.

The first half is roughly nine minutes long, which in a full show usually gives enough time for a couple more vertical manoeuvres, such as a Quarter Clover or the challenging Swan to Fred loop transition: challenging because the formation are at their furthest distance from the leader while in Swan and, as they fly the loop, they collectively move forward into the Fred position, which is tightly packed around the leader. Then add to this a couple of rolls, the hardest being in Chevron or Big Vixen. In between these sequences the jets come back in to form other shapes, such as the Concorde or Blackbird, or wider shapes like the Phoenix, Lancaster or Spitfire. Key to all Red Arrows displays is, of course, the trademark Diamond Nine formation, which is flown either as a short Diamond or a show Diamond.

During the first half of a display Red 1 will always try to keep the show 'loaded' – meaning that they are pulling G-force through acceleration, vertical manoeuvring and sharp turns. In a loop, you will pull about 3–4G on the way up and 4G at the bottom. While pilots must work against the physical effect of G-force through breathing and tensing techniques, having an aircraft 'under load' (i.e. under constant G-force) helps pilots to keep their position within a formation, resulting in tight formations without any quivering wings.

Flat displays are much harder to fly, principally because you're coming on and off G all the time, as a result of doing lots of level turns. With only 1,000 feet of sky to work with – a letterbox, compared to the completely blue canvas that the pilots have to play with in a full show – there is no space to do loops or rolls, and there are even greater demands on your flying technique and accuracy than there would be in a full show. No one in the team enjoys what we call the 'flattie'; they would much rather be looping and rolling. And yet there is always satisfaction to be had in delivering a show despite the great British weather working against us, with the crowds beneath us absolutely drenched in rain.

In the second half of the show, which is usually ten minutes long, Reds 6, 7, 8 and 9 come into their own. Red 1 and the rest of the Enid section are given a little more time offstage and handover to the rear section, round the corner at the edge of the display site before coming back to the main display area. This is a section full of G-force, with about twelve to fourteen separate handovers between Reds 1 and 6 (the rear section leader), and timing is incredibly important. Both of their cockpit stopwatches will be in constant use, as every manoeuvre or contract handover is timed to the second. Typical manoeuvres include the rear section pass, which consists of the pairs 6 and 7, plus 8 and 9 performing an opposition pass, hurtling towards each other at 100–300 feet above the ground and crossing each other at an eye-watering combined speed of 700mph. The Carousel manoeuvre really tests the synchro pair as they perform opposition passes in front of the crowd, then turn and perform another further back, before finishing off in a circle at the front. Another popular manoeuvre, known as Goose, sees Red 8 or 9 sprinting towards the five aircraft of Enid in Pyramid formation and crossing straight through the middle of them – this move, in particular, requires almost superhuman levels of flying skill and nerve.

The Mirror Flat and Mirror Roll are manoeuvres performed by the synchro pair, together with Reds 8 and 9. As they fly in a vertical column, the synchro pair (Reds 6 and 7) invert themselves so that they are directly above Reds 8 and 9, with the canopies of their jets just thirty feet apart. To hold his positioning reference while inverted, Red 7 has to push himself back into his seat and crane his neck to look up at Red 6 – the two aircraft being so close that Red 7 can almost see the writing on the kneeboard of Red 6. Maintaining your position while inverted is doubly tricky, as the control stick moves in the opposite way: nudge the stick left and the jet will move right, and vice versa – it's an incredibly skilful manoeuvre.

A real crowd-pleaser and signature manoeuvre for the synchro pair is the Heart and Spear, which involves the two aircraft performing a vertical pull-up, then separating and rolling with the smoke on, to create the top part of the Heart shape. Roll too fast and they have lost the central dip; and if they are too close, they will end up with a little hole and more of a potato shape (and the synchro pair never wants to be told at the debrief that they flew a Potato rather than a Heart). The Heart makes for a popular shape, as it can of course be dedicated to certain events, such as Valentine's Day; and in April 2020, Reds 6 and 7 flew a blue-shaped heart over RAF Scampton to show their appreciation for the NHS and key workers during the Covid-19 pandemic.

Usually, while Red 9 flies the Spear for the Heart, Red 8 might be performing elsewhere in something like the Goose (there are versions in which one or two aircraft 'goose' the Pyramid). In the second half of the display Reds 8 and 9 flit about in separate manoeuvres, moving between sections as safely and as dynamically as possible, before they find each other, join either Enid or the rear section and then split again. The flexibility of these two roles was something that appealed to former Red 1 Martin Pert, who flew as Red 8 in 2014: 'I really enjoyed flying as Red 8. You're out of formation more, and you pride yourself on how quickly you can get to other sections. One minute you're doing a loop with the front section to go into a split and then, thirty seconds later, you're with another section doing a mirror manoeuvre. No one on the ground realises it's the same aircraft that's just done the loop.'

During the second half the Enid section will often perform Roll-backs, which involve two of the aircraft pulling up, rolling around 360 degrees and slotting back into formation, before another two jets do the same and repeat. It's a challenging manoeuvre and one of the hardest ones to train for and learn. Another popular manoeuvre with the team is the Python in which Enid flies two big barrel rolls, one after another, covering six miles of ground-track – this is something I always loved flying in displays. To sign off the show with something impressive, the team goes for a final break, something like the Vixen Break, which sees Enid with Reds 8 and 9 in Line Abreast, charging towards the crowd before they all fan out with a perfect plume of red, white and blue smoke. This is always a great way to provide a final huzzah to the show, and the crowd roars as Red 10 signs off his commentary on the ground with: 'You have been watching the Red Arrows!'

HAPPY LANDINGS

While the show may now be over, the team must return to base or move on to their next display or fly-past, still working to the minute-by-minute schedule of the WHAM. Transits to and from shows often prove more complex than the displays themselves – the team having practised and perfected their choreography for months – and landing while in formation can be challenging. With around five miles to go, Red 1 will need to line up the full formation (in Battle V) with the runway, and he or she has to narrate every minor turn over the radio – a tiny bit left or a tiny bit right – which can sometimes go on a bit, much to the amusement of the team.

Once the team is perfectly lined up, they will descend to about 500 feet, at which point the levels of concentration and focus are really high amongst the team. Red 1 will call, 'Reds *breaking* now!' and on the b of '*breaking*', all pilots turn their smoke off, before peeling out of formation in perfect two-second intervals. Then it's wheels down, flaps down and tip-in, ready for touchdown. Everyone aims to touch down at between 2,000–4,000-foot spacing, and if there's wind from left or right we will take that into account and modify the way we turn our final corner to land.

A really spectacular landing that we sometimes go for is the Spaghetti Break. We'll run in in Diamond formation, before climbing into a loop. Then, as we get to the top of the loop, we'll split the formation so that we're pointing directly at the ground from about 6,000 feet. Red 1 will call, 'Break, break, go!' and we'll all turn our jets onto our individual pre-briefed angles and fan out, leaving a cascade of white smoke trailing down to as low as 1,000 feet. Then everyone will turn their smoke off and we'll safely come into the landing pattern.

What I especially love about the Spaghetti Break is that we will sometimes do it at a normal regional airport, such as Exeter, where there will be people simply climbing the stairs onto their commercial flight. They're not at a display, they're just off on holiday, and it's such a surprise when they see the Red Arrows perform some wizardry in the sky before we come in to land. To see something as awesome as that, when you're not expecting it – well, I like to think it would make an excellent start to anyone's holiday.

8

The Blues

BEFORE A PILOT STRAPS into a Red Arrows jet, there are engineers on the ground who have already primed that Hawk so that it performs perfectly in the sky: obediently answering the slightest pressure on its control stick, producing just as much thrust as is needed, while withstanding the rigours of high-G manoeuvres. Ahead of a day's events, the Red Arrows' operational support staff will have ensured that airspace is cleared and safety regulations have been met, with the whole squadron working to a complex schedule that guarantees those nine jets arrive where and when they should, with the famous smoke ready to go.

The personnel behind all of this, who keep the Red Arrows flying and operating throughout the year, are a 115-strong dedicated ground crew. They range from engineers and other technicians, to drivers, photographers, supply specialists, a public-relations section and other vital support-staff members of RAFAT. Named after the colour of the flying suits they wear, the Blues are the powerhouse of the Red Arrows machine. The largest section of the Blues consists of engineers. They comprise eighty-five of the total support team, because keeping a minimum of nine fast jets in prime condition throughout the year requires round-the-clock work by skilled technicians. During the winter months the jets require more substantial work, with in-depth servicing and system checks, while during the summer maintenance and repairs must be turned around quickly, if the team is to keep up with its busy schedule of displays and events.

The Blues are made up of a wide assortment of professions and specialist skills, providing a great overview of the immense variety seen right across the RAF. These include mechanical, avionics-weapons and survival-equipment technicians, and supply teams who transport the thousands of pieces of equipment needed by the team to wherever they are in the world. Across the squadron, personnel range from young apprentices on the first rung of their RAF career to those who have spent more than thirty years in the military. They include regular servicemen and women, full-time reserve personnel, civil servants and a small number of contractors. This is known as the Whole Force and it is a very efficient, diverse and highly motivated team. As Officer Commanding, I have oversight of the entire squadron and am responsible for the welfare of the 125 personnel who make up RAFAT, and I make sure the team is adequately trained, that our output is safe and that communication between all the various sections is effective. The management of the Blues is split between various department heads, who, alongside me, Red 1 and Red 10, form RAFAT's senior leadership team.

JENGO AND SENGO

For the engineering sections of the Blues, this departmental leadership falls to the Senior and Junior Engineering Officers (known as SEngO and JEngO respectively), who are both commissioned officers. The JEngO manages the day-to-day output of the unit, ensuring that pilots have sufficient aircraft for their needs, whether it's for training or for events in the display season. The SEngO, meanwhile, strikes a balance between overseeing more of the long-term planning needs of the unit's aircraft and making time-critical airworthiness decisions relating to all the complex problems that arise in ensuring safe fast-jet operations. The SEngO leads and works alongside the rest of the management team, which includes another junior officer as well as a more experienced Warrant Officer and Flight Sergeant.

In 2019, Flt Lt Ben Ireland (since promoted) took over from Alicia 'Lissy' Mason as the Junior Engineering Officer for the team. It was a job that he had set his sights on some nine years earlier, when he joined the Royal Air Force. With experience working on fast jets, in particular the Tornado GR4 before it went out of service, Ben was invited to apply to RAFAT, joining the team in June 2019. Having steered his career towards the Red Arrows, he was thrilled to take up the role and was ready to tackle the particular demands of the job: 'As an engineer, I've always been obsessed with problem-solving and am keen to work in lots of different environments with the Air Force. With the Red Arrows, you often find it's just you and your small team in some random airfield in the UK or abroad, and if there's a problem, it's up to you to solve it without the support or tooling you'd have back at base. It's a huge challenge, and that aspect of the job really appealed to me.'

Much of the work of the JEngO involves dealing with any faults or emergencies that arise, meaning that a quiet day can suddenly ramp up to being a very busy and pressurised one. As a Flight Commander, he would also manage the careers and well-being of his team, who work round the clock – day and night – with a typical day's shift made up of around thirty to forty people who attend to (or, in RAF parlance, 'spanner') the aircraft. During winter training those on the day-shift start at 7 a.m. and, after an engineering briefing, Ben and his team would look at each aircraft and assess which ones are serviceable. He would then head up to the main weather brief with the pilots and other senior team members, before informing them how many aircraft he can supply for training – the objective being that they always have more than the team needs each day, while also balancing the extensive servicing needs of some of the jets in the fleet. Ben would also make himself available during the day when pilots sign the jets back in after sorties, so that they could communicate with him any issues or problems they might have experienced with the aircraft.

Set against the requirements of training is the need to give some of the aircraft extensive overhauls, and a typical winter's day might see two jets in the hangar having a 'primary star service', during which technicians and engineers run through a number of inspections and set tasks. Occasionally an aircraft will be stripped back entirely, with all its parts removed, so that it looks like an empty shell; this enables its systems and parts to be reconditioned, checked and put back together – a huge undertaking, which can take four to five months. The painting of aircraft, so that they retain their famous red sheen, is also undertaken during the winter months, the objective now being that each jet is painted every five years. With the process taking about a month, Ben must juggle how to achieve that alongside all the other requirements of the team.

Despite the many demands of the job, working on the team's jets is something of a pleasure for engineers like Ben – the BAE Systems Hawk T1 jet used by RAFAT is a mechanically driven aircraft, very different from the fly-by-wire computer-led systems of the Typhoons and now fifth-generation fighters. Some of the Hawks used by the Red Arrows were first issued to the team back in the 1980s. In fact as recently as 2018 a jet that made up part of the Red Arrows' original Hawk fleet for the 1980 season was still being used by the team (tail number XX227). For all the team's Hawks, the cockpit design is very analogue: dials and switches, rather than screens and displays. 'When you take off the panels and look under the skin of the aircraft,' Ben explains, 'you can see everything and, more often than not, you can physically see what's wrong with the aircraft. With a more modern aircraft, such as the Typhoon, a good two-thirds of faults are down to computer issues, not faults you can actually *see*. For an engineer, working on

Hawks – some of which are a good ten years older than me – is a great place to be.'

While many of the techniques and tools used by the Blues are the same as they were forty-odd years ago, the engineering team has brought in many modern updates to help manage the aircraft, including new testing techniques and the use of computational methods to monitor the structural health of the aircraft, which is carried out by specialist RAF engineering teams. When it comes to taking an engine out and replacing it with another one, however, the engineers employ much the same skills and methods that have been used over the past four decades.

THE CIRCUS

A distinctly unusual – and much-coveted – role for the JEngO is that of 'Circus 1', who flies in the back seat of Red 1's aircraft during the display season as part of the travelling support team. Another nine engineering technicians and one photographer make up the full Circus team, and occasionally SEngO will fly with me in Red 11, as – you guessed it – Circus 11. Each Circus engineer is allocated to a specific pilot throughout the season, flying to and from display airfields and servicing the aircraft before and after every show. It's a system that has always worked well for the Reds, having been introduced in 1967 when the team was flying a fleet of Gnats. Without the Circus engineers on board, ready and primed to work at every destination, the team simply could not achieve its vast number of displays in the UK and further afield in any given season.

The Circus is generally selected each December, with the Junior and Senior Engineering Officers, plus the Warrant Officer and Flight Sergeant, drawing up a shortlist of recommended technicians for me to approve. Although they must choose individuals who are technically competent and able to cope with the weighty responsibilities of the work, they also need to consider the character and experience of each member. Trust binds everyone at RAFAT, and pilots and Circus members are in each other's company for a large part of every day during the season. In our high-pressured environment, it's vital that they get along and communicate well. The Engineering Officers will consider a range of factors when allocating Circus members to pilots: a really competent but young engineer might be matched with a more experienced, easy-going pilot, in a bid to ensure that the two get along and work well, in what is the fairly confined space inside a two-seater Hawk.

Factor in displays across the UK, together with long-haul transits across the Atlantic or to the other side of the world, and it's not surprising that a real bond develops between pilots and their respective Circus members. It's a connection that can last down the years, with former team leaders Ian Dick and Dickie Duckett – both Red 1s in the 1970s – still being in touch with their respective Circus members. They recall how engineers had just as much (if not more) affection for the individual aircraft they maintained as the pilots did, often claiming proprietary rights over the aircraft, which pilots merely 'borrowed' from them from time to time. 'They were worth their weight in gold,' says Ian Dick of the Circus. 'It was incredible what they did, in terms of looking after the aircraft. I had complete and utter faith in them.'

On long-haul transits flying at a high altitude there is often plenty of time for pilots to chat with their Circus member, as I remember doing with (then) Flt Lt Marcus Ramsden on the tour to China, and with Lissy Mason in my later years as Red 1. Up above those clouds, with the Earth far below, you're in a bit of a rarefied, almost serene bubble, with just air-traffic control in your ear, plus the voice of whoever is in your back seat.

Travelling together over long and arduous trips can strengthen your alliance, and the Circus engineer may witness at first hand exactly how difficult conditions can get for the pilot in the front, as Ben Ireland remembers, when flying with team leader Martin Pert in the summer of 2020: 'We did a fly-past over London with our counterparts from the French Air Force. The weather was bad, we were on a really strict timeline and I knew this wasn't the time to be conversational, so I sat really quietly. I could tell that Perty, who has enormous mental capacity and is normally really composed, was having to work extremely hard, as I could hear how heavily he was breathing – it is an intimate connection.'

In the Royal Air Force it is rare for engineers to fly in a fast jet on a regular basis, and as Circus members fly in the Hawk whenever the team is moving from one location to another, they can end up amassing hundreds of flying hours. Before joining the team in the back seat, each Circus engineer must undergo a medical test, during which measurements will be taken to ensure they can fit in the small cockpit and reach all the necessary controls. As with the pilots, they also do swimming tests, undergo ejection-seat training and sea-survival training, in case they should land in water. This involves being dragged along by boats, so that they learn how to detach their parachutes and haul themselves into their life raft, which is contained in their personal survival pack. Just like the Red Arrows pilots, the Circus engineers must also learn to recognise the symptoms of hypoxia (oxygen deprivation) and be prepared for the physical stress of sitting in the back of a fast jet. They also undergo high-G training at RAF Cranwell, where they are taught how to do anti-G straining methods and are spun around in a rig at a high rate of G.

With all the basic survival training in place, Circus members then get to experience what they call a 'Shakedown' flight, when they are taken up for the first time in a Hawk. To prepare them for this, the Circus undergo a simulator visit with Red 10, sit in a dummy ejection seat and attend familiarisation briefings, usually with their respective pilot. The subsequent Shakedown flight is tailor-made to expose the Circus to the operational and physiological realities of fast-jet aviation. Prior to this, many of the engineers will have had very little flying experience, other than the odd flight in a commercial airliner. As a result, that first flight in the Hawk makes for a memorable (if challenging) experience for the Circus engineers. Simply getting into all the flying and safety gear, breathing the pressurised air and experiencing all that G-force – not to mention the adrenaline that comes with it – means that most of them are left exhausted for the rest of the day. Some might feel a little unwell or even vomit, but most agree that it's an exhilarating experience, even if they are also rather pleased to be back on the ground after that first flight.

However, all Circus engineers ultimately get used to the sensations of flying in the Hawk and count themselves lucky to be able to do so. The Reds usually transit between airfields at a low level, between 1,000 and 2,000 feet, and at that altitude you really feel the speed of the aircraft. 'It's incredible seeing the landscape zip by,' says Ben Ireland, 'and at that height you can pick out the details on the ground – cars, people and farm animals – and it really is a pleasurable experience.' The Circus engineers don't have too long to take in the view, as transit times are pretty quick in a jet flying at 350–400mph.

The role of the Circus also is to 'see off' the aircraft before any sortie, which involves a set number of final procedures to check that the aircraft are working as they should. Engineers will meet their respective pilot at their aircraft, exchange a few words and help their pilots strap into their seat. Once the engine is running and the pilots are going through their own start-up procedures, the engineers will then do a further check for fuel and hydraulic leaks. During the display season, when the Reds are in front of spectators, the Circus performs a 'formal start', in which the see-off procedures and checks are synchronised, to add a little more theatricality to the proceedings. Standing in a long line in front of their respective jets, they will do all their checks – such as for air brakes and then flaps – and, after each check, they will give a hand-signal, look to whoever is standing in the middle of the line, put their hands by their sides and, once the person in the middle gives the nod, they will move on to the next part of the check. Safety is still the prime objective in the formal start, but making it more coordinated gives it a little more flair, as befits the UK's premier display formation team.

SMOKE AND ARMOURERS

The Red Arrows are known for their colourful vapour trails (often called 'smoke'), which provide a distinctive visual impact to fly-pasts and displays. In flying the flag for the UK, the red, white and blue trails provide a heart-stirring spectacle. For trips overseas the team occasionally honours the colours of other nations, as they have done for Greece during our training period there, with blue and white trails. The popular Heart shape, amongst other manoeuvres, can only be created with the use of vapour trails, which serve to enhance the drama and impact of many of the team's manoeuvres during displays. The trails are also vital for safety, enabling pilots to judge wind speed and direction, and to spot each other if the formation has split into sections and is several miles apart. Aircraft will rotate around contour trails in certain manoeuvres, and the vapour trails are critical in head-to-head passes. As a result, white smoke is used throughout training, with the red and blue colours added during Springhawk in the build-up to the display season.

Essential to the Blues, therefore, is the dedicated dye team, which ensures that each aircraft is able to trail the distinctive vapour as soon as Red 1 makes the call, 'Smoke on, go!' The white vapour is produced by injecting diesel into the exhaust of the jet engine, which, at more than 400 degrees Celsius, vaporises immediately. The blue and red colours are made by mixing dye with the diesel, both of which are stored in pods fitted to each aircraft, and each colour is released by the pilot pushing one of three buttons. Usually a pilot has five minutes' worth of white smoke, and just one minute each of red and blue smoke, so it's critical that the team makes precise use of the smoke it has, working to a pre-planned 'smoke plot' for every display. To ensure an environmentally sustainable smoke-generation system for the future, an exciting project is under way with academia and industry to redesign our delivery of this colourful and safety-critical element of the Red Arrows display.

During the display season the dye team travels by road so that it can replenish the pods ready for each display, and often works to a tight timescale. The process involves pumping in diesel followed by a dye/diesel mixture, with each display using one barrel of red dye and one of blue dye, topped up with diesel. As there is no dial or gauge to indicate whether the pod is full, instead the dye-team technicians must crouch down and put their ear to the tank – once they hear a change of tone, they know the pod is full. As the dye colours stain, the technicians must wear protective overalls, gloves and goggles, which can make for sweltering work when it's a hot summer's day, or if the Reds have just touched down in the Middle East.

Amongst the Blues there are also weapons technicians – known colloquially as 'armourers' – who maintain the aircraft's ejection seats, explosive cockpit glass canopies and fire-suppression systems. Should a pilot or Circus member need to initiate an ejection, he or she must pull a handle between their legs. One-tenth of a second afterwards, an explosive detonating cord (which runs through the glass canopy in a jagged arrow-pattern overhead) shatters the canopy, and then more explosives drive the seat upwards along telescopic rails. Rockets thrust the seat a further 200 feet up into the air and the parachute opens. To prevent inadvertent firing of the ejector seat, the Red Arrow pilot, prior to any sortie, will visually check that the safety pins are in their correct positions.

Once all occupants are correctly strapped into the aircraft, the canopy is lowered and locked shut, prior to engine start. Before the start sequence is activated, each occupant of the ejection seat must, in a slow and deliberate manner, move two pins to make the seat 'live' – one pin activates the canopy detonation cord, and the other seat-firing pin is stowed between the occupant's legs, directly underneath the yellow and black pull-handle that initiates the ejection sequence. Both pins are moved into their 'stowage' area just above the pilot's left shoulder, where it is visible to the ground crew, who also conduct a visual check confirming whether the seat is 'live' or not. When the aircraft requires maintenance on the ejection seat, the armourers will move additional 'safe for maintenance' pins, which shut down the whole operation of the ejection system.

BEHIND THE CURTAIN

Alongside the engineering team are other support staff at RAF Scampton, including sections led by the Team Manager, currently Squadron Leader Doug Smith, the Communications Director Andrew Morton and the Team Adjutant, Warrant Officer Alan Irons. Each of their respective teams plays a vital, often behind-the-scenes role in ensuring that plans run smoothly. This is particularly critical during the busy summer season. In a typical year the Red Arrows carry out, on average, more than seventy displays across four and a half months. August is the busiest month, during which the team performs publicly five days a week, with a typical day consisting of multiple transits, air shows, fly-pasts and ground activities.

The Reds regularly display at the major national air shows, such as the RIAT, and at coastal events, such as the Bournemouth Air Festival – with the latter drawing in more than a million spectators over a single weekend. The team will also do on average six bespoke high-profile fly-pasts, such as the Queen's Birthday Fly-past in June or the Great North Run in September, alongside around 100 fly-pasts en route to other scheduled events. These might be anything from a fly-past over an event or festival with a crowd of thousands, to a small village fete – so long as it's an open public event and no more than around fifteen nautical miles away from their transit route, the Red Arrows can usually put on a bit of smoke and streak across the skies to help bring the event to life. The domestic events also dovetail with an extensive international schedule of air shows and fly-pasts in mainland Europe or further afield, and the team builds in at least one major overseas tour every other year.

Running this vast operation of events is the support team, which controls the logistical, financial, brand and administrative aspects of the Red Arrows squadron while working closely with sponsors, the media, government departments, the MOD and the Royal Air Force chain of command. By the end of October the RAF's air-events team at RAF Northolt will have received bids from show organisers and the public who have requested displays and fly-pasts, a rough draft of which will initially be sent to Red 10 – Squadron Leader Adam Collins from 2018 to 2022, with Squadron Leader Graeme Muscat taking over the role in March 2022. Once he or she has received the drafted requests, Red 10 will set about building a display-season plan for the summer. The aim is to ensure that the display sites are safe and appropriate for a Red Arrows display, all the while liaising with the Civil Aviation Authority (CAA) and the military equivalent (the Military Aviation Authority), which set the regulations and rules for display flying.

As a nine-ship formation, the Red Arrows take up a lot of ground-track and far more airspace than a single aircraft or Typhoon doing tight manoeuvres. Added to this, the synchro pair flies as low as 100 feet above the ground, so Red 10 must check that the relevant clearances are in place to enable the team to operate safely. Red 10 will visit any new display sites, map them and write a comprehensive report of his findings for further review, before presenting his findings to me, as Officer Commanding. We will then go through each display-site report in detail, before presenting the risk mitigations to our chain of command. Red 10 will also liaise with the Air Crew Planning Officer, who will help with the logistics of the displays, working out transit times and routes for the team, building in contingencies for bad weather, or time-lags and turns if the team is running early. Once the plans are approved, the support team takes over, as they build up the complex and myriad logistics of a busy display season – logistics so complex that you might often hear the engineers say that the pilots have it easy: they simply turn up and fly the aircraft.

The team also has a manager who, among other things, spends a large part of his or her time planning the ground logistics of overseas tours, making sure that the Red Arrows team is where it needs to be and when, while the Adjutant – a military officer who gives administrative assistance to a senior officer – will have overseen other arrangements, including booking transport to and from various airfields and researching and arranging accommodation. The Communications Director – whose remit includes media opportunities and PR – will have worked with event organisers and other key partners to ensure that ground engagements are well planned and meet the overall aims of the visit. And the Team Manager or the Operations Officer (usually an experienced Flight Lieutenant) will oversee the logistics of getting support staff and equipment to locations overseas, in tandem with the Red Arrows' engineering sections and with specialist, highly-trained RAF units. The team may be allocated an RAF transport aircraft, such as an Airbus A400M Atlas, to take everything it needs to a certain location, while other items might need to be sent on ahead on a container ship. This all needs to work alongside the team's requirements in the UK: a dye-pumping machine, which enables the Reds to trail their colourful smoke, might need to be sent out on a ship, meaning that the team has one less machine at home, limiting what the team can do domestically. Planning – and sometimes expectation management – is vital.

The Operations Officer helps to plan flying and associated activity, both in the future and on a day-to-day basis back in the UK, and produces the document that pilots, engineers and the whole team refers to as the WHAM: 'What Happens According to Mange' – a nod to this document having been, historically, overseen by the Team Manager. This document schedules, in minute detail, the hours and minutes of the pilots, engineers and support staff: every arrival and departure, allowing at least twenty minutes for briefings and another fifteen minutes for the check-in before take-off. Engineers know where they should be at any given minute, as do the dye team, PR teams and other support members; everything is itemised, planned and scheduled, so that nine aircraft arrive at a prescribed time.

The WHAM also details the transport of bags, something that might seem rather trivial, but is crucial for the smooth running of the team. Alongside their flying responsibilities, team pilots also have secondary duties, which might include rations (food) or ensuring that bags are delivered when and where they should be. These might seem menial jobs for seasoned fighter pilots, but in line with our team philosophy, everyone plays his or her part in making sure each task is planned and executed correctly. Each of the ten Hawks has a small two- by three-foot space for bags, called a 'pannier', and because three of these need to contain a spare wheel, a jack and a toolkit, that leaves just seven panniers for twenty people, so packing light is encouraged. Anything put in a pannier needs to be minimal (though we do occasionally get road support when bigger events demand that we carry extra bags). Team members need to think ahead and pack carefully because the space in their pannier bags, which already contain at least one change of flying suit (out of the three that pilots are allocated), is at a premium.

Alongside their flying responsibilities, the Red Arrows team also attends a large number of ground events that fulfil a variety of roles, from showcasing the skills of the RAF and aiding recruitment to the armed forces, to supporting British industry and attending various STEM and educational events. They also take time to meet the team's many fans, plus members of the media. One weekend might see the Reds appearing across the UK – at Prestwick in Scotland, Portrush in Northern Ireland and Torbay in England – or the team might put on a smaller display at somewhere like Peterhead in the north-east of Scotland, especially if there is a particular focus that draws them to the area, such as recruitment or supporting an event.

Beyond the air shows, the Red Arrows also appear at more general events like the Goodwood Festival of Speed, which showcases a variety of fast machines and British engineering. There we will have pilots, engineers and other staff on the ground talking to the public, and will help support a variety of high-performing and inspiring brands and manufacturers. It's an extraordinarily diverse schedule of events, with the team tailoring their promotional responsibilities accordingly. 'In terms of tempo and profile, it's a bit like running a combination of a Formula 1 team, an aerobatic team and a rock band all at the same time,' says Comms Director Andrew Morton, 'but, quite rightly, within the setting of what is a UK military team and an RAF squadron.'

CAPTURING THE MOMENT

Critical to the team is Red 10, who, alongside his work of liaising with display organisers, flies the tenth aircraft in the team with a photographer in his back seat, supervises the training and safety aspects and provides the commentary at displays. It's an incredibly varied and responsible position, requiring both flying and supervisory expertise, and is a role that Squadron Leader Adam Collins has performed from 2018 to 2022. Having flown the Tornado GR4, the F-111 (on exchange with the Royal Australian Air Force) and the Hawk T1 as a Flight Commander with 100 Squadron, Adam had the necessary fast-jet experience and airmanship skills that are demanded of a Red 10. Up until 1997 the job of Team Manager and Red 10 were one and the same, but the increased safety and supervisory requirements of the team necessitated a splitting of the roles. Red 10 flies in formation with the team when it is transiting, often tucked behind Red 1, if they're in Battle formation.

The longest-serving Red 10 – indeed, the longest-serving team member in the Red Arrows – is Mike Ling, who took up the role in December 2011, having recovered from our accident in Crete in 2010 and following three years with the team, flying as Reds 3, 7 and 6. As Red 10, Mike filled in for the many prominent fly-pasts the team did in 2012 when, after the death of Flt Lt Sean Cunningham and the departure of Flt Lt Kirsty Murphy, we had to fly as a seven-ship formation in displays. Mike was originally meant to stay for two years, but continued in the role for another six years, returning briefly to the team in 2018 to fill in as Red 3.

Mike comments, 'I almost enjoyed being Red 10 more than being in the team. It's classed as a desk job, but you amass a couple of hundred hours a year in a jet and flying in a helicopter to more remote display locations. During training you give instant feedback, so you are also changing the performance level of the team.' Throughout the year Red 10 oversees safety and performance from the ground, including the more dynamic moves of the synchro pair, radioing through feedback to section leaders as they fly and contributing to the all-important debriefs. Red 10's main role as a pilot is to fly with a photographer in his back seat, positioning his Hawk alongside the formation so that the photographer can get the perfect shot of the team in action, but without putting anyone in danger. It's a demanding task, particularly when the formation is performing looping manoeuvres down to 300 feet. 'Whilst you want to get a good shot,' explains Adam, 'you need also to be really aware of your positioning. If someone needs to break out of formation, I need to be sure I'm not in the way.'

The team has three photographers, and between them they film training and public displays from the ground. For safety reasons, all practice sorties and displays are filmed from the ground – not from the air – and are analysed in the debrief afterwards. One of the photographers also usually flies in the back seat of Red 10's aircraft, as Circus 10, to and from events. Photographs and video footage of the Red Arrows shared on social and mainstream media have always offered a fantastic opportunity to showcase the team, capturing everything from when the nine gleaming jets are lined up at an airfield, to the Reds in action and performing an awesome variety of manoeuvres: upside-down, rolling around smoke or executing a spectacular break to finish off a show. Since 1965 the Red Arrows have also been captured soaring over major landmarks around the word, from Edinburgh Castle in Scotland to the skyscrapers in Chicago, from the ancient ruins of Petra in Jordan to a fly-past over the Petronas Twin Towers in Kuala Lumpur, Malaysia. It all provides an amazing photo opportunity, and numerous images of the Reds remain displayed long afterwards in embassies, high commissions, government buildings, schools and universities across the world.

To achieve that fantastic imagery, Red 10 will carefully plan and discuss with the photographer prior to getting airborne the types of images they want to capture, perhaps focusing on a particular manoeuvre or section so that they know which position to get into, to line up the perfect shot. Experienced as Red 10 and Circus 10 are, everything from weather conditions to how the light falls (not to mention a little luck) can play its part in the process of capturing the moment – occurrences that Red 10 must constantly react to. 'Sometimes you just have to see where the sun and landmarks are when you're up there,' explains Adam. 'Although the shots are carefully planned, you don't know quite how it will look until you are in position.' While shooting stills in the back seat, the photographer also has a GoPro strapped to the lens of his or her SLR digital camera, so that there is always running footage of the team in action. Capturing the team flying over an incredible cityscape, or as part of a unique formation with another big aircraft, is a definite perk of the job, and Mike Ling maintains that 'the best part of being Red 10 is photo-chasing. You really do have the best seat in the house.'

Alongside the event admin, photo-chasing and safety supervision, Red 10 also provides on-ground commentary for the displays, while also liaising with Red 1 on the radio and checking the site for any unforeseen safety issues. A typical display day might involve Red 10 transiting in formation with the rest of the team and landing with them at a local airfield. While the Circus services the jets, Red 10 and his photographer quickly jump into a Central Flying School helicopter to make the fifteen-minute journey to the show site, avoiding roads, which are often clogged up with large arriving crowds. Red 10 will aim to arrive at a display site a good ninety minutes before the start of the show, giving him enough time to speak to the display organisers and to radio Red 1 with the wind and pressure readings and details of onsite weather conditions.

Red 10 will also scan the area for any unforeseen hazards or structures that might not have featured on the mapping documents that were prepared for the team ahead of their arrival. Nine times out of ten, it's adverse weather that causes problems, but Red 10 will need to be on the lookout for anything hazardous – from increased bird activity (particularly if the site is a seaside location), to something that might have strayed into the display area, such as a hot-air balloon or a ship with a tall mast. Temporary structures that have popped up, such as a Ferris wheel or a large marquee, might not be of any real danger to the team, but Red 10 may give Red 1 a heads-up about it all the same, stating its position relative to the show centre.

As Reds 1 to 9 head to the display site, Red 10 – as the only red suit on the ground – might do some media interviews or talk to the public, before giving Red 1 any last-minute information on the radio. He or she will then begin commentating to the crowds, which could number anything from a couple of hundred people to close to one million, if it's one of the bigger air shows. Red 10 will alert spectators to the Reds' imminent arrival, adding snippets of information about the pilots and their RAF careers, while narrating and describing the manoeuvres and shapes that the team performs. With the microphone in one hand and the radio to Red 1 in the other, Red 10 will occasionally hold both of them close together so that the crowd can hear the leader's radio calls in real time.

Giving the commentary can take a little getting used to, and although there is some media training, the best way to get comfortable with it is simply to do it – as with flying, practice makes perfect. 'I didn't like doing the commentary at first,' Mike Ling remarks (which came as something of a surprise to me, as I know him to be quite a talkative and engaging character). 'I just didn't enjoy it. Then I found a rhythm and realised I could play on the crowd's reactions. I then embraced it, and by the end I thoroughly enjoyed it.' On the rare occasions when there's a delay, Red 10 needs to think quickly on his feet, which sometimes gives him the opportunity to inject a bit of fun, as Mike also remembers: 'Towards the end of 2012 the guys were late for a Weymouth display, so I ended up getting the whole beach doing a Mexican wave. I got into the groove and decided not to use any scripted jokes, because there are lots of repeat customers at air shows and I don't want to reel out the same old puns.'

Once the show is over – the crowd hopefully rather more jubilant now than when they arrived – Red 10 and the photographer will jump back into the helicopter to join the rest of the team for a debriefing, using the video footage taken from the ground; or, if there's another event scheduled, Red 10 will head off with the photographer to the next site, to repeat the process. It's a well-worn and polished procedure, which allows the Reds to move from site to site efficiently and safely, enabling the team to perform at two or three shows in a single day. If large numbers of people were hauling in truckloads of equipment to every show, the Reds would be less nimble, and less capable of putting on multiple twenty-minute shows to a broad variety of audiences across the country. With Red 10 and his photographer zipping between sites, the Circus at hand to engineer the Hawks, and the support staff busily planning the next move, the whole operation is just as precise and streamlined as a Red Arrows performance in the sky.

9

Risk

THE RED ARROWS aerobatic displays are impressive for the glorious, heart-stirring spectacle of nine gleaming jets streaking through the sky, but also for their tight formations, knife-edge turns and seemingly impossible manoeuvres, which to those on the ground look like a short ride to oblivion. Just how dangerous is it to be a Red Arrow, how does the team manage the risk and what do they do if the unexpected happens?

The Red Arrows' *Display Directive*, a thickly bound document that is commonly referred to as 'the bible' of the team, provides all the answers. It describes the operating procedures of the Red Arrows team, and the objectives are clearly outlined: in enhancing and protecting the RAF's reputation for excellence, the aim is to conduct 'a safe, memorable, visually exciting formation display', and that order of priority is clear. An exhaustive tome, the *Directive* covers everything from the safety kit that pilots must wear, to G-force procedures and the training required for Circus engineers to fly in the back seats of the Hawks. The identification of risk, and how it can be reduced or avoided, is a central thread running throughout – every procedure of the Red Arrows being built around the core principle of safety.

The mitigation of risk begins with the selection of some of the RAF's most able and skilled pilots for the team, who then embark on the rigorous and incremental training programme. The team will work through manoeuvres and shapes, ensuring that it can fly each element safely and competently before moving on to another one, building up those cognitive processes so that they are almost hard-wired into the brain. In building sequences that require inch-perfect precision, the team must always try to factor in the unexpected and will rehearse a range of emergency scenarios in simulators at RAF Valley, ranging from radio transmitter (RT) failure to an engine malfunction or a bird strike. Briefings before every sortie will highlight potential hazards, taking into account the wind and weather, and expected G-force load. The leader will, at random, also choose a team member to go through at least one emergency scenario, and they need to demonstrate that they know the relevant radio call and how to remove themselves from the formation safely if needed. Debriefings after every sortie will be similarly rigorous, with the leader and pilots highlighting any mistakes or safety hazards that may have been experienced during the flight, which must be talked through and resolved prior to the next sortie.

Also outlined in the *Directive* are the display sequences and manoeuvres that the team intends to fly in the upcoming season, as authorised via the awarding of the Public Display Authority (PDA). Granted annually at the end of Springhawk training, the PDA involves detailed scrutiny of activities both on the ground and in the air and is awarded only if the senior chain of command is convinced that the team has a show that is safe and presentable. The *Directive* outlines every formation shape, manoeuvre and synchronised pass with descriptions, techniques, leader's calls, what to do in an emergency or RT failure along with the relevant escape manoeuvres. All formation positions and manoeuvres have prescribed escape procedures, if a pilot cannot maintain his or her formation position or sight references, perhaps because they are blinded by a winter sun or there is a loose object in the cockpit. For the majority of formation positions there is a basic escape manoeuvre, which is to pull upwards and out of the formation until fully separated from the other jets. All new and revised manoeuvres are thoroughly assessed in terms of risk, and from all angles, such as: what is the impact of G-force; what is the aircraft manoeuvrability envelope (outside which a manoeuvre is dangerous); and what are the potential hazards?

The Military Aviation Authority (MAA) is the regulating body of all defence aviation activities, including the Red Arrows team, and all the displays are operated in strict accordance with UK and global display regulations. At any display, the safety of the public is paramount. For each public display, the team will run a risk assessment and it's the job of Red 10 to liaise with display organisers to ensure that they are compliant with the relevant regulations. The leader will review all the safety aspects on the ground, present any potential hazards to me, as Officer Commanding, and we will grade what those risk elements might be. For example, if it's a coastal site there needs to be a marine exclusion zone to stop boats coming in, to avoid the kind of scenario seen in 1980 when a Red Arrow clipped a boat mast in Brighton. We will also consider whether the site has a busy airspace around it, such as Biggin Hill, which is close to Gatwick airport. We'll also look at the time of day a display takes place: if we're displaying late in the afternoon or evening might there be an increased bird-strike risk?

If a display site is deemed high risk, then the decision will be not to attend. The Shoreham Air Show disaster of August 2015, during which a former military aircraft crashed, hitting vehicles on the nearby A27 road, killing eleven people and injuring another sixteen, led to a tightening of regulations for air displays. Due to the high speed and dynamic nature of Red Arrows displays, the RAF conducted a separate assessment of the risk associated with Farnborough International Air Show – at which the Red Arrows have displayed most years since the team was formed – and concluded that it was no longer possible for the team to perform an aerobatic display there, due to the increased amount of housing and businesses built by developers adjacent to the airfield along with major transport links. Having had a long association with Farnborough, the team is still keen to support the air show as tasked by the RAF much as possible, so instead the Red Arrows have subsequently performed various fly-pasts and have conducted a range of ground engagements, including STEM and engineering talks, rather than full displays.

Any accident, such as the Shoreham Air Show disaster, sends shockwaves through all display teams and aviation units, prompting the authorities to re-evaluate the risk and regulations and to take on board any lessons that can be learned. Outside the UK, air shows around the world have seen some terrible crashes: in 2002 the Sknyliv Air Show in Ukraine saw an aircraft plough into the crowd, killing seventy-seven people and injuring nearly 300, while in Germany, the Ramstein Air Show collision in 1988 saw the death of three pilots from the Italian Air Force display team and of sixty-seven spectators, with hundreds of others injured. We are particularly sensitive to tragedies experienced by other formation aerobatic teams – most recently, the crash in 2020 of a Canadian Air Force Snowbird jet, which led to the death of one of its team members. There are so few aerobatic-formation teams around the world (many of whom we know and have flown with) that accidents of this nature are felt by everyone in the Red Arrows team.

Over its fifty-eight years, the Red Arrows have lost ten pilots and one Circus engineer, tragedies that have had a profound effect on the team, and which the squadron and the RAF strive exhaustively to prevent. Often these tragic incidents act as a reminder that flying – of whatever nature – always comes with an element of risk, as all RAF pilots are acutely aware, which is why those who join the team practise their displays so relentlessly and to perfection. Throughout the history of the Red Arrows, the team has safely performed almost 5,000 displays, has transited thousands of miles across the world, and has conducted many thousands (if not millions) of hours of training without incident. On those few occasions when there has been an incident, most have resulted in the damage or loss of aircraft, with pilots managing to land or eject safely – possibly being injured in the process, but alive. The team's first fatality occurred in its fifth display season, in 1969, when Flt Lt Jeremy Bowler flew into trees during a practice. Thereafter the team's worst accident occurred in 1971 when the synchro pair struck each other while practising the Roulette manoeuvre at 100 feet above the ground, causing both aircraft to crash, killing four pilots – the two current team members and two new team pilots who were flying in their back seats. A tragedy of this scale led some people to question whether certain manoeuvres were simply too dangerous to perform. At subsequent air shows the public expressed more appreciation and affection for the team than ever before, with the tragedies laying bare the skill, precision and courage required of the Red Arrows.

THE RISKS OF FLYING

That sympathy and connection with the public were things I experienced first-hand when, on 20 August 2011, my very good friend and colleague Flt Lt Jon Egging – known as 'Eggman' – died in a crash towards the end of his first year flying as Red 4.

At the time, I was nearing the end of my first three-year tour with the Red Arrows. That day we had been deployed to a regular fixture on the RAFAT calendar, the Bournemouth Air Festival where we flew a display and two local fly-pasts. At around 1.45 in the afternoon we headed back to Bournemouth airport, flying in Big Vixen, one of our core triangular-shaped formations, with Reds 1 to 5 forming the V-shape; as Red 6, I was flying with Red 7 at the rear. As we approached the airport, Red 1 radioed to the team to commence a 'flat left/right break to land', so rather than pulling into a loop prior to the break to land, the formation initiates a straight and level run, then pairs of aircraft –first Reds 8 and 9, followed by 4 and 5, then 2 and 3, 6 and 7, and finally 1 – pull outwards and upwards, climbing 500 feet in order to decelerate the aircraft. Once the call was given, Reds 8 and 9 rolled and pitched outwards, followed by Red 4 (Jon Egging) and Red 5. Jon verbally initiated the break (the last transmission made by him), but halfway into his turn he started to overbank, with his aircraft nose increasingly pointing downwards. Red 2, noticing that Jon was descending, immediately transmitted two radio warnings – '4, check height' – without any response. Seconds afterwards, I saw Jon's aircraft crash into open fields near the bankside of the River Stour, and Jon was subsequently declared dead at the scene.

The rest of the team landed, knowing that Jon had crashed, but we hadn't had the confirmation that he had been killed. The systems controlling what happens after a major accident or incident immediately kicked in, and the whole team huddled in a crew room at Bournemouth airport. There, we wrote down everything we could remember about the lead-up to the crash – the idea being to get our thoughts down in their purest form before they became skewed or unwittingly biased further down the line; these thoughts would form part of the written statements used for the crash investigation. Very quickly, crash-management specialists arrived on the scene to make sure the site was inspected and protected and all evidence was gathered in line with regulations. For me and the team, the whole formality of the process helped, especially when we still didn't know whether Jon had died, as it required us to keep a professional mindset and focus, as best we could, on what had just happened and to do what was right, both in response to the accident and for Jon.

After that we simply sat tight, still at Bournemouth airport and in a state of shock, having been informed that Jon had died, but unable to phone our loved ones, which was something we were all desperate to do. It was becoming evident that news of the crash was already spreading on social media. However, across the military, the protocol is that the next of kin must be informed of any fatality or serious injury before anyone else makes contact with the outside world. It's a system designed to protect whoever is to receive the worst news. The priority was to make sure that Jon's wife, Dr Emma Egging, was informed with the most up-to-date and correct information. The partners of team members rarely travel to the display locations; however, this weekend Emma was at the Bournemouth Air Festival in the company of close friends. Afterwards the team began grappling with the grief of losing a close friend and colleague. To get over that initial shock, our natural instinct was to huddle together and talk about as much as we could collectively, while also allowing anyone who needed to remove themselves from the group to do so – although we might shadow or check in on them frequently to make sure they were okay.

I'll always remember how phenomenal the public response to the crash was. The following morning the team walked down to the Town Hall in Bournemouth, where streams of people were already lining up to sign first two, then four and finally six books of condolence set up by the local council in honour of Jon. I can still picture all the flowers, candles and notes – full of messages such as 'You filled our skies with colour and our hearts with joy and excitement' – and there was an overwhelming amount of love, warmth and support from the public.

With the immediate start of an independent investigation into the accident, decisions needed to made by the senior chain of command as to what the team should do in the short term: should we be entirely grounded or continue in some form? From my own and the team's perspective, there seemed no obvious explanation for the accident. The day had passed smoothly without incident, the conditions and visibility were considered good and there appeared to be no issues with Jon's aircraft or his handling of the jet. I knew Jon very well, having trained and worked with him a number of years previously as a Hawk T1 instructor, before he went on to fly Harriers on the front line. He was an exceptional pilot, had made an outstanding start to his tour at the Red Arrows and had flown a break-to-land on an almost daily basis since becoming a fast-jet pilot – and, prior to the flight, Jon was fit, well and in his usual good humour.

The inquiry thoroughly assessed a whole range of factors that could have led to the accident. Inspection of the aircraft wreckage concluded that neither technical nor engineering failure of the aircraft had led to the accident, nor had there been any attempt to eject, prior to impact. It also ruled out the possibility of negligence or sabotage, and there was nothing in Jon's medical history to indicate abnormality. The inquiry eventually concluded that the most likely cause for the crash was a loss of consciousness during the break, as a result of 'A-loc', 'Almost loss of consciousness', a condition whereby an individual suffers profound cognitive and functional impairment after experiencing G-force. In total, the inquiry panel made forty-two recommendations, including: an improved high-G training syllabus across the RAF; ensuring adequate maintenance and repair procedures to all anti-G equipment; and a complete reassessment of the team's methodology for examining all of their aviation and engineering risks. Jon was well versed in dealing with G; in fact I had always described him as a bit of a 'G-monster': he had a short, stocky frame with good muscle build, all of which helps to combat G. On the many times that we flew aerobatics together I would sometimes climb out of my aircraft totally done in by the G-force, practically dragging my knuckles on the ground, while Jon would almost leap out of his aircraft. Overall, the fact that this tragic event could happen to Jon underlined for us the critical necessity of implementing the inquiry recommendations as soon as possible, to ensure that all of our G-force training, awareness and measures were as good as they could possibly be to prevent this scenario from ever happening again.

The senior chain of command eventually established that it was safe for us to get back in the air a couple of weeks after the accident, and to fly the jets back to Scampton from Bournemouth. Balanced against any incident is the need to get the team back in their aircraft quickly, otherwise their cognitive processes and flying skills can erode. The day that we first strapped back in was, of course, difficult; none of us had been near those jets since the accident and the flight back required some serious compartmentalisation, putting aside the emotions that we all felt, so that we could focus on the task in front of us. Before that day, more than twenty years had passed since the Red Arrows had suffered any fatalities – the last being in 1988 when Flt Lt Neil MacLachlan was killed during a practice at the team's base at RAF Scampton. And before that, the last fatal accident had occurred in 1978 when former team leader Wing Commander Dennis Hazell was killed alongside Flt Lt Stephen Noble as they performed a Rollback.

It was decided that we would try to finish the rest of the season as an eight-ship formation. Considering that Jon had died on 20 August and the season was due to finish in early October, that was quite a challenge, as we had to reorganise ourselves very quickly. We approached the revision of the show very gradually, and at briefings it was important that we looked Red 1 in the eyes and told him straight whether or not we were fit to fly. I think we were all experienced and empowered enough to know that if we didn't psychologically feel up to it, given everything that had happened, we could say so. What helped was that we talked through and identified every little thing that could affect us once we were in the aircraft, pre-arming ourselves for anything that might throw us. We were all feeling Jon's loss, and we knew that the first check-in – when we run through all the Red numbers in order – would be a particularly poignant one. Ahead of time, we talked about whether the boss would say the number 4 or whether we would jump from 3 to 5, so that we were mentally and emotionally prepared for the moment.

We pulled together to restart the display season, and delivering the first show after the accident safely and to a high standard felt particularly meaningful to the team. Throughout the rest of the season everyone talked about Jon. People thanked us for carrying on and that really helped us to feel – in the face of all the emotional and practical difficulties – that keeping going had ultimately been the best thing for the team to do. Life for Jon's family was of course incredibly difficult, particularly as the formal inquests for such events tend to linger on for months or even years, with the inquiry into his death reaching its final conclusion in December the following year. The accident was also covered extensively in the media, making any form of closure impossible for family members.

Within twenty-four hours of his death, I remember Jon's widow, Emma, saying that she wanted to do something in honour of him, as Jon was such an inspiring individual. In his memory she set up the Jon Egging Trust, which has supported thousands of vulnerable young people over the years, helping them to reach their full potential by giving them access to inspirational training and opportunities linked to the world of aviation.

Having completed the season, I left the team in early October 2011 and had a couple of months to train and prepare for a front-line tour in Afghanistan. On leaving the team, and away from the day-to-day challenges of flying with the Reds, some of the emotions about Jon's death came to the surface – I was clearly still processing much of what had happened in August. Hearing about the death of Flt Lt Sean Cunningham just a month later, on 8 November, was another shock that left me numb: I couldn't imagine the grief his family was going through, and how the team was going to cope with a second fatality in such a short space of time.

Sean had been with the team since October 2010, flying first as Red 3 and, by the time of his accident, having just completed five weeks of winter training as Red 5. South African-born and brought up in the UK, he had flown Tornados on the front line, including operational sorties in Iraq, was a tactics and pilot instructor and was regarded as one of the RAF's most skilled aviators. On the morning of 8 November Sean had strapped himself into his aircraft as he prepared for a work-up sortie with Enid. Squadron Leader Jim Turner, who had taken over as team leader only three weeks before, describes the incident: 'There were five aircraft parked on the pan, and I remember it being a misty morning in November and, as the weather wasn't great, our plan was to fly to RAF Valley. I'd just told everyone to start their engines, we had engineers in our back seats, and there was this big explosion. Someone on the radio said that Sean had ejected and I saw his seat being fired up out of his aircraft, but his main parachute didn't open and he impacted the ground.' Still attached to his seat, Sean had fallen 220 feet and was killed by the force of impact.

Again the team was grounded, along with all Hawk variants in the RAF and any aircraft, such as the Tucano or Tornado, fitted with Mk10 ejection seats. The team was grounded for five weeks and, with two fatalities in the space of four months (not to mention my collision with Mike Ling in 2010), team leader Jim Turner was concerned that the future of the Red Arrows was decidedly precarious, although he was quickly assured by the Chief of Air Staff that the Reds would get through this. The team did manage to resume training, although not properly until February 2012, and was subject to frequent safety inspections and audits. Securing PDA in April, the team continued as a seven-ship formation, bringing in Dave Davies and Mike Ling to make up the nine-ship formation for the many special fly-pasts that the team flew that Jubilee and Olympic year.

When the team flew as a seven-ship formation, Mike Ling was on the ground commentating for the summer displays and was once again bowled over by the response from the public: 'The reaction of the crowds to the seven-ship display that year was noticeably different. People wanted to talk to me about Jon and Sean, and I was really quite surprised by the level of esteem and warmth there was. In my commentary I spoke about Jon and Sean and dedicated part of a manoeuvre to them, which made for a really moving moment.'

While the team pressed on with a busy summer schedule, investigations into Sean's accident continued, the principal questions being why had Sean ejected when he had no need to, and why the ejection seat had not worked properly. The process, however, would last much longer, as both military and then civil inquests examined the evidence, much of which centred on the mechanism of the ejection seat. The inquests concluded that the seat's firing-mechanism handle had been left in an unsafe position, which would have facilitated an accidental initiation of the ejection seat with a forward movement of the hand. The handle had probably been pulled by the aircraft's seat straps in a previous sortie and it hadn't been noticed by anyone, as the seat's safety pin could still be inserted. Once the ejection seat was initiated, the parachute then failed to deploy because a nut and bolt had been fastened too tightly – a potential hazard that the manufacturer, Martin-Baker, had identified, but not effectively communicated to the RAF, although it wouldn't be until February 2018 that the company accepted that it had breached safety laws.

The inquest ultimately proved that the fault with the ejection seat could have caused the death of any RAF pilot who sat in the Hawk T1, regardless of whether or not he or she was a Red Arrows pilot. The accident also highlighted how reliant pilots are on the thousands of different elements and mechanisms that make up the lump of metal they take up into the air – along with the correct supply, fitting and regular maintenance of those elements so that they operate as they should, whether on the ground or flying at 400mph. It's a hugely complex operation, and the reliability of the team's Hawks and the relative rarity of accidents are a testament to the skill of our engineers and operational staff at the Red Arrows and the wider RAF.

The accidents of 2011 prompted a reassessment of various supervisory structures and processes at RAFAT, while also providing a stark reminder of the inherent risks of flying. I, perhaps more than some others, was conscious of those risks, having experienced a mid-air collision only the previous year – an incident that, in the space of just a few seconds, could have spelled the end of both me and my synchro partner, Mike Ling.

A CLOSE CALL

By March 2010 I was in my second year with the team, this time in the Red 7 position in the synchro pair, having been selected by synchro leader Red 6, Mike Ling ('Lingy'). As the name suggests, the synchro pair performs the synchronised opposition and head-to-head passes, involving low-level rolls and sustained high-G turns. Mike and I had been training from the October of the previous year in a gradual learning process, with Red 6 coaching Red 7 at every stage. As Red 3 in Enid, I had never been required to perform below 300 feet, so my first task as Red 7 was to learn how to manoeuvre the aircraft safely 100 feet above the ground. Then, once Red 6, Red 1 and the Officer Commanding were content that I had safely mastered that, I moved on to the opposition or head-to-head passes. This involves getting used to, and building up, a mental image of what it's like to head towards another jet – a dot in the distance – which suddenly flashes past at a closure speed of 700–800mph. On the practice flights you begin with a wide distance between the two aircraft and then gradually move in, until you have reached the minimum lateral separation of 100 feet – it might look like mere inches when you're viewing it from the ground, but 100 feet between you and another aircraft is actually a close, yet still relatively comfortable distance.

Once that mental image is set and you feel comfortable doing the head-to-head pass, you then build up the various manoeuvres that require an opposition pass. We had completed all the manoeuvres and only had five weeks to go until PDA, the plan being that the team would train at Kastelli Air Base in Crete for the first two weeks of Springhawk and then move to RAF Akrotiri in Cyprus for the assessment. When we arrived in Crete, on 21 March, both Mike and I had felt that the work-up had gone really well – there didn't seem to be any moves that we felt we couldn't do or had been particularly troublesome. So at that stage we were finessing it all, putting on the final polish and making the most of those cloudless skies of the eastern Mediterranean.

The crash that ensued happened on the second day of training in Crete. Some accidents occur when people are feeling over-stretched, perhaps if they're pushing themselves a bit too hard or there are other distractions at play. But in many regards this wasn't the case with us – we both felt comfortable and sure that we'd cracked the training programme; nor did we feel compla-cent, but rather focused on getting our turns and rolls really crisp. We both felt supported by everyone around us, the jets were in good order, the weather was great and we were well rested. And yet, as you saw in the *Prologue*, we went from every-thing slotting into place perfectly to almost killing each other within a second.

As I mentioned earlier, Mike and I were busy rehearsing the opposition barrel roll, a manoeuvre that required us to cross our flight paths prior to doing the opposition pass and we knew we had a minimum distance at which we could safely cross our flight paths. After we had both made the 'threshold' radio transmission to confirm that we were happy to enter the manoeuvre, we both suddenly realised that the view ahead of us was different – the aircraft opposite looked much bigger and closer than it should have. Faced with the unexpected, I had to decide immediately what to do next: should I enact what, ultimately, I should have done, which was the prescribed escape manoeuvre (one of many that we set in place for every part of the display), or should I try a different method of escape? In that split second, seeing Mike's Hawk racing towards me, I felt that the escape manoeuvre – which was to roll the jet away and then pull away – would actually put myself in more danger, as I might end up putting the belly of my aircraft into Mike's Hawk.

So (and the mental processing time really was incredibly short) I decided that the best thing to do in that moment was to push forward on the control column and try to avoid Mike by going down rather than up; to fly beneath him instead of rolling away, my brain thinking, 'If I can see him the whole time, I'll miss him.' I had moved from a 'conditioned response' (the escape man-oeuvre that we go over again and again in briefings) to an 'instinct response', and the unfortunate fact was that Mike had also had the very same instinct response. Seeing him also push down, and our Hawks getting closer and closer, I realised that we had no chance: I knew we were going to collide, and I thought to myself, 'Right, there's no get-out here – I'm done.' I ducked my head and shut my eyes, and waited for the inevitable.

Then I heard what I can only describe as a deep thud, causing the airframe to vibrate momentarily and shudder. I opened my eyes and thought, 'Okay, I'm still here.' Then suddenly all I could see was the runway hurtling towards me as my jet headed straight down towards it, the collision having happened just 110 feet above the ground. Rather than ride my jet straight into the ground, my brain, fuelled with adrenaline, kicked into rapid response. With every sinew of my body on red alert, I pulled back on the control column and my Hawk lurched up and away.

However, Mike was dealing with his own nightmare situation: 'Monty's tail collided with my wing, and the tip of his tailfin hit my canopy about eighteen inches above my head. There was a big bang, lots of wind noise as we were travelling at three hun-dred and fifty mph, and shattered Perspex in my face.' With his aircraft heading downwards, Mike instantly pulled on the ejec-tion handle, seconds before his jet smashed into the ground. His Hawk splintered on impact, littering its burning parts across the airfield below.

Meanwhile I was flying away from the scene, in obvious shock, but within about fifteen seconds of the collision, my brain – in some kind of manic response mode – had entirely planned how Mike and I were going to get back in the air and everything would be okay. Those thoughts would evaporate as soon as I saw in my mirror the explosion from Mike's jet hitting the ground. There followed the worst couple of minutes of my life, when all I could conclude was that Mike had been killed. Red 11, who was on the ground, then called through on the radio in a panicked voice, 'STOP STOP STOP!' – which is the call that is made when there's an incident and the entire display must immediately be abandoned. At the time of the collision the rest of the formation was airborne, during what was the third practice sortie of the day; and Red 1, Ben Murphy, had seen our accident as he came round ready for their next manoeuvre.

Still believing that Lingy had died, I tried to slow down my breathing and compartmentalise, with an internal monologue of 'Right, Lingy has now gone, but you are still here and you need to land safely somehow.' Red 11 then called through, 'Shoot seen' (as in a parachute had been seen), indicating that Lingy had ejected. But in my super-heightened state, I misinterpreted it as 'No shoot seen', and my worst fears were seemingly confirmed: 'Okay, Lingy has gone, deal with it later, just get this aircraft down.' I was overwhelmed with crushing guilt – 'I've done this, I am responsible' – emotions that I desperately tried to push away. 'Just fly the plane, Monty.'

Feeling very much alone, convinced that I'd killed my synchro leader, I suddenly saw another Hawk come and join me on my wing. It was Red 9, New Zealander Simon 'Kermit' Rea, who had been instructed by the leader to come and give me a bit of support. I was still feeling pretty desperate, but it was a comfort having him there and I remember looking over and thinking, 'At least I've got Kermit.' At the same time I seemed to be flying the Hawk as if on automatic pilot, dealing with the required hand–eye coordination to fly the aircraft, although I hadn't consciously realised at the time that it kept wanting to roll to one side. After another few minutes I heard a second call from Red 11 on the radio, and this time there was no mistaking his words: 'I see him moving.' To my utter relief, I realised that Lingy had got out of his aircraft – he was alive.

Lingy had in fact lost consciousness when his head hit the back of his seat during the ejection, but had come to when he hit the ground. 'The first thing I remember was the smell of smoke and burning and I couldn't work out if I'd ejected or not,' remembers Mike. 'I was metres away from the fireball and burning wreckage and I thought: "Have I just hit the ground in the aircraft and survived it?" What gave it away was seeing the cord next to me that attaches the survival pack with the lifejacket and, once I'd spotted that, I knew I had ejected.

'The next thing I did was sit up, as I wanted to know where Monty was. I was in a lot of pain, as my shoulder was dislocated, and people were on the scene within seconds. When I asked about Monty, they told me not to worry as he was still airborne. I could only think they were lying to me and I remember saying, "He can't be, we've just flown into each other." It was only when I heard Monty's voice on a hand-held radio that someone had that I knew he was okay.'

Safe in the knowledge that Lingy was alive, I focused on what was going on with my aircraft. Kermit, on my wing, gave me a report of how my Hawk looked from his vantage point, delivered in true Kiwi style, casual-as-you-like: 'Oh, mate, it's not looking good. You've lost a bit [of this] and a bit [of that] – I can't lie, it's looking pretty shabby.' I can't quite remember his exact words, but they were uttered in such a run-of-the-mill way that I couldn't help but laugh – that was exactly what I needed, and it served to take the edge off my shredded nerves.

Thinking a little more clearly by now, I went over in my mind what I should do and through the processes of checking the aircraft, at which point I realised that it felt like it wanted to lean left quite considerably. Kermit had said that I had lost part of my fin and, unbeknownst to me, my tailplane was damaged, as it had of course sliced through Lingy's canopy. However, I still felt I could control the jet, so I decided not to eject – or not yet at least. Where to land was my next conundrum, as I couldn't land at Kastelli airfield because it was now a crash site, so instead I needed to divert to Heraklion international airport. That in itself required a fair amount of mental agility, figuring out which heading and route to take while talking to air-traffic control, but somehow I managed to get myself, and Kermit, to Heraklion airport.

As I had control issues with my Hawk, I couldn't simply land at the airport. I needed to run a mandatory assessment, known as a Low-Speed Handling Check, to confirm that if I took it down to the runway, it wouldn't tip over and crash at the last minute. As a result, my plan was to get out over the coastline of Crete and run the check over the sea, so that if I did lose control I could eject and the aircraft would simply plunge into the waters below – hopefully without me in it. As I ran through the test, I therefore had to get ready for an ejection, checking the various straps and going through the mentally gruelling prospect of potentially launching myself into the sky, strapped to a rocket-fired missile.

Luckily it didn't come to that. I decided that the aircraft was sufficiently okay to head towards the runway, and I managed to land safely. As soon as I touched down, I realised I wasn't getting any feedback from the rudder at my feet, so I'd had no lateral control of the aircraft while airborne. If there had been a strong cross-wind when I approached the runway, I would have tracked sideways, unable to use lateral control to bring the aircraft round and put it on the ground. Thankfully there wasn't much of a cross-wind that day, so I could use a bit of a roll and get the jet down. Kermit landed after me and, once we had taxied to a standstill (amid all the other usual airliners that you find at an international airport) and our engines had stopped, I stepped out of my aircraft, took a few deep breaths and had a good look at my jet.

I was surprised at how beaten up it looked and I hadn't realised how much of the tailplane had gone: the fin at the top had ripped off, leaving a ragged edge, and the rudder was only loosely connected to the tailfin assembly and fell off by the time the engineering team got to the aircraft. Considering how many bits of metal were missing, I was amazed that the aircraft had flown as well as it had. As I inspected my Hawk, Kermit came over, gave me a hug and I asked him how Lingy was, but he had only as much information as I did. I then made a quick call to my partner. She had already seen footage of the collision – a freelance reporter had captured the incident on a mobile phone and within twenty minutes of impact, the whole thing had been shown on the mainstream media.

About fifteen minutes later the rest of the team landed at Heraklion and I remember saying 'I'm so sorry' to Red 1. The enormity of the guilt was starting to take root in my mind: I knew this incident would put the whole team seriously off-track. In Ben Murphy's typically assured and calm manner, he told me not to worry, that they would figure out what to do and at that point we needed to concentrate on Lingy.

Although he was now conscious, Lingy was in a bad way. The force of the ejection – at such high speed, he had no time to position his body correctly – had knocked him unconscious and dislocated his shoulder. He also had an injured leg, a fractured hand and lacerations all over his face, caused by the explosion of his Perspex canopy. The ejection mechanism, however, had done exactly what it was meant to and had ultimately saved his life. He was immediately taken off to hospital, undergoing a fairly gruelling ambulance ride along the winding, mountainous roads of Crete. Reds 1, 9, 11 and I were sent back to a hotel and, in the usual evidence-gathering process, wrote down everything we could remember about the accident.

Later on we headed to the hospital to visit Lingy, with the team giving me five minutes alone with him before they all came into the room. The bond you have with your synchro partner is particularly close: for months you have trained together every day, building the trust that the partnership relies upon, and for that to go wrong was a massive shock. There he lay in a hospital bed – bruised and battered, with cuts all over his face – whereas I barely had a scratch on me. I struggled again with that enormous sense of guilt. As the rest of the team joined us, I attempted to maintain a brave face, keeping my feelings very much on the inside;

but during that hour with him, my heart sank more and more as I thought, 'Look what I've done to Lingy.'

AFTERMATH

The day after the collision I was interviewed by an investigation board, which informed me that Lingy and I were to be flown out of Heraklion on an RAF medical flight two days later. The RAF has exceptional recovery systems in place, which means it can pick up injured or deceased servicemen and women pretty much anywhere in the world and get them quickly back to the UK. We ended up on a VC10 with a whole squadron of engineers and pilots returning from an exercise in Oman, and as Mike had also damaged his lungs, the transporter needed to fly at a lower level than normal. 'I was strapped down and just elated to be alive,' he remembers, 'although I was quite blinkered to the situation and thought I might be back in a cockpit in a couple of weeks.'

Unlike Mike, I was completely fine physically and, being acutely aware of the serious casualties that were routinely flown aboard this aircraft, I kept telling the medics that I really was okay. Lingy was sent to a military trauma ward in Birmingham, where he was told that the injury to his leg was serious, requiring surgery and quite a few months' recovery. For Mike it was 'gutting news' but, he recalls, 'I couldn't feel too sorry for myself. It was the height of the wars in Iraq and Afghanistan and I was in a military ward in Selly Oak Hospital and every other person in that ward had a life-changing injury incurred on operations. That really put everything into perspective for me and got me through those weeks.'

With Lingy in hospital, I was back at home, and after two or three days I got a call from Red 1. Having established how I was, he got straight to the point. 'I need to ask you one question: are you prepared to get back in the air asap?'

While at a basic level I was fine, the thought of strapping myself back into that Hawk any time soon felt pretty daunting, so

I answered, 'In all honesty, I'm not entirely sure.' It wasn't quite the answer he was looking for – indecision is not a trait that is admired in the military – and he pressed me for a yes or no, understandably needing to know how to proceed with the team in Crete. I mulled it over a little more, still unsure what to do, until I suddenly came to the decision while driving my car. I pulled over and called Red 1, sitting by the side of a road in north London, and told him: *Yes*.

In all honesty, I was only about 80 per cent sure that I wanted to get back in the air, but I felt it was enough to go on. However, there was still that niggling thought: 'Can I really do this?' The senior chain of command had obviously made their initial investigations into the accident and concluded that it was safe for the rest of the team to progress with training, if it could. Lingy was likely to be out for a minimum of seven months, and the team had already identified a previous Red Arrow who had flown in the synchro pair, Paul 'Pablo' O'Grady. He was flying Typhoons at the time, but it was hoped that his chain of command would be willing to release him, given the circumstances. The option was either to cancel the synchro pair or bring Pablo back. The latter happened, Pablo joined the team, and I was now facing the prospect of getting back in a Hawk.

As a result, three and a half weeks after the impact, with the rest of the team still overseas on Springhawk, I was back at Scampton, where it was decided that Pablo and I would try to build up the show again, but this time without the opposition barrel manoeuvre. The investigation had already identified that the main cause of the accident was the inherent risk involved in crossing flight paths prior to doing the barrel roll. The conclusion was that simply passing each other, without crossing flight paths, provided the same visual result from the ground but was much safer – and that is how the manoeuvre is shown to this day.

Lingy was still in hospital and had received the unwelcome news, before he was discharged, that his other leg was also damaged, so he would need to go through additional treatment. He consequently had to use a wheelchair for three months, with the

RAF Benevolent Fund loaning him an electric chair, as he couldn't push himself around on his recently dislocated shoulder. Meanwhile I tried to grapple with a ton of what-ifs. What if I had a flashback or suddenly froze in mid-air? Back then (and we'd probably do things a bit differently now) the psychological assessment took a common-sense approach, with senior officers looking me in the eye and asking whether I was okay. Pablo also constantly checked in on me – was I good to go? – and I always knew that if I was really struggling, I could say no.

Initially I flew two or three solo flights, with someone in the back seat, just in case I had any kind of nervous collapse in the cockpit. It felt fine, and with every sortie I did I felt increasingly positive. I felt like I was rectifying something of what I'd done to Mike, and I was desperate not to let the team down. It wasn't all roses, though: I wasn't sleeping well and was starting to have a few minor flashbacks; but on the whole I generally felt pretty balanced and kept a check on myself as we went along. I made sure that I felt safe taking off, flying and landing, and a gradual approach – solo flights, followed eventually by the more aerobatic manoeuvres – was a crucial part of that.

A key manoeuvre to crack was, of course, the opposition pass: would my nerves hold, knowing full well what could go wrong? By then Mike had joined us at Scampton and he was a real moral support, reassuring me the whole way. He played a big part in getting me back to flying like I used to. The time pressure was also immense, as we had only three weeks to build up an entire show; to get through it, Pablo and I formed a kind of bubble with Mike, Red 11 David Firth-Wigglesworth, and our Group Captain. For the first opposition pass, Pablo had designed a very basic sortie to reduce the pressure and, as we taxied away, I remember waving to Lingy, who was out on the runway in his wheelchair, and I thought, 'Christ, look what happened to him the last time I did this.' Thankfully, I kept my unhelpful thoughts in check and the sortie went okay. Very slowly, with Pablo checking on me after every flight, we built up the complexity over time and managed to work up the synchro-pair elements of the show within the tight three-week window.

However, a few weeks later when I was back with the team overseas and we were getting the whole display together, the stress of the collision started to surface a bit more. I had a really bad period of sleeping, and I was feeling uncomfortable – when flying past Pablo in various opposition passes, I would feel a shot of unease. This emotion was new to me, something that hadn't occurred when getting back into the cockpit after the crash or at any other time as part of the synchro pair. I tried to suppress it, but I could tell that a form of anxiety was building.

REPLAY

Looking back on this period, I realise that my psychological state was fairly fragile. My greatest counsellors had always been the team. I knew that I had to talk about the events of 23 March to somebody, and I found that I was able to talk openly about the collision with team members without becoming emotional, and I felt this was a good foundation. However, I hadn't discussed my growing anxiety with anyone. When I returned to the UK immediately after the crash, I was scheduled to meet the Senior Medical Officer on base. We had a very quick and frank discussion, and he looked directly at me and asked how my mental state was. At the time I genuinely thought I was okay and answered honestly, 'I'm all right.' After a few moments of consideration, he nodded and reminded me that if I needed professional counselling, it was available. In many ways I really appreciated how pragmatic our exchange had been, but this was only the beginning, instead of a conclusion.

A week or so after the crash the investigation team came to see me, this time with a psychologist, in order to piece together the cognitive aspects of the crash. Their schedule was busy, so they couldn't get to Scampton until early evening. We talked for hours, but the turning point in the conversation came when I was shown the video recording of the crash for the first time. They asked me a series of questions about my interpretation of what had happened moments before impact and, on showing me the tape, awaited my reaction. Suffice to say, watching the recording was awful. It was the first time I had seen the complete after-effects post-impact: the fireball and the remnants of Lingy's parachute as he landed painfully on the ground. From the air I had seen the smoke and the crash site, but I hadn't witnessed the enormity of Lingy's jet bursting into flames.

In watching the tape, I relived the emotional response on the day of the accident, a rich mixture of guilt and relief. And in some ways the worst part of it was that I immediately defaulted to type – I began to discuss the entire event as though I were in a RAFAT debrief, yet without the psychological safety net of the team around me. I mused that my transition to the line-up for the manoeuvre wasn't as good as it normally had been, and I could see a slight correction in my flight path. I remembered doing this in the moment, but was still appalled to see it on the tape. Most notable, though, was how my brain seemed to have understood already that our positioning wasn't right, but my reaction to this information wasn't immediate. I could almost see the lag-time, with my brain frantically processing the situation. The telltale sign was that a specific radio call (made prior to the manoeuvre) should have coincided with a change in our smoke colour. But rather than changing my smoke from white to red, I had stayed on white – a mistake I had never made before. I'm sure that my brain was busy thinking about why things didn't look right, instead of going into its conditioned response of changing the smoke colour and continuing into the manoeuvre. I openly berated myself in front of the investigation team for not having responded differently, analysing every second where I had missed an opportunity to prevent such a catastrophic outcome.

We turned the tape off, sat in silence for a few seconds and then I asked for a two-minute break before the questions resumed. I wasn't emotional, but I had an overriding sense of disappointment – in myself and in the entire situation. Of course I rationalised that

Lingy and I had a huge amount to be grateful for: we were both still alive, for a start. But I think this episode cemented my overwhelming sense of responsibility for the crash, for the state Lingy was in and for the whole reputation of the team. Looking back, it was after this round of questioning that I most needed help, but I didn't seek it and, by unfortunate coincidence, no one else was around later that evening. The team was in Cyprus (having moved on from Crete), my other half was in London and I remember feeling incredibly low when I returned home very late that night.

The investigation team and the psychologist had been friendly and considerate, but they were under their own pressure to gather as much information as possible, and this meeting had given them a lot. For me, it had also taken a lot. I got back to an empty house and reached for the whisky. I'm not a regular whisky drinker, saving it mostly for good times with friends, but on this occasion I indulged in the clichéd response of drowning my sorrows. With my forehead pressed firmly on the kitchen table, I resolved in my mind that, no matter what, I wouldn't let the team down again.

I have no doubt that internalising this decision deterred me from talking openly about some of my anxieties, although had they continued, I am sure I would have raised my hand. However, on returning to Cyprus a strange episode occurred, which in some ways started to turn things around for me. I had always taken much pride in the fact that I had never missed a morning alarm clock or a single important 'get up' in my RAF career. And yet on the visit to Cyprus of our Air Officer Commanding, Air Vice-Marshal Mark Green, to see how we were getting on after the crash, I inexplicably missed my wake-up alarm. When I was not seen in the mess at breakfast or during the short drive to work, the Operations Assistant, Corporal Pete Jones, was tasked to get me out of bed asap, which he did by rapping his knuckles loudly on my door. After getting over the sickening feeling that I had missed the first sortie of the day, I realised that I had six or seven minutes to get there, so I dry-shaved, whipped on my flying suit and raced to work. I made it, with a minute to go, before the sortie brief.

Squadron Leader Graham Duff, Red 8, gave me a wry smile as I scurried onto the outside benches, our usual morning-briefing location. Duffy, who for me was something of a paternal-cum-eccentric uncle-figure, asked if I was okay and offered me a chocolate bar – 'breakfast', he indicated. I gratefully accepted the sugar-hit, and took a moment to reflect on whether I should go flying. I decided to stay with it and enter the briefing cycle with the team, with the AOC looking on. As the brief went on, I became less and less foggy and, after correctly answering a question on an emergency scenario, I decided I was good to go. Pablo walked closely by me on the way out to the jets, seeking one final confirmation that I was fit to fly. By now the adrenaline had kicked in and I answered honestly that I was okay – I had overslept, but I was mentally ready and fit to fly.

The sortie went well. In fact it went really well and was one of my better performances of the week. Incredulous that I had missed my alarm, I focused purely on making things right, and none of my previous anxiety played out. The next two sorties that day were also fine, and I started to feel better about things. In fact from that day on – and I'm not sure why, although I certainly never slept in again – I felt steadily more comfortable back in the cockpit, pointing my jet at my synchro leader's oncoming Hawk at a closure speed of 800mph. After a few more weeks we eventually secured the PDA, and the focus on delivering the season took hold for all of us.

In truth, the subsequent public shows provided the biggest boost to my confidence and mindset – we were showered with support and affection, and I remember getting lots of hugs from people who expressed how relieved they were that Mike and I were okay. The success of the season, and the positive sentiments we received from the public, really validated our decision to carry on. That twitchy period I had in Cyprus – well, it was definitely worth it.

Of course Lingy couldn't fly that season, and that huge regret never left me, but we involved him as much as possible and he

24. A Red Arrows team photo, spaced so as to be covid-compliant, taken during Springhawk 2021 in Tanagra Air Base, Greece. The team pilots wear their famous red flying suits, with the ground crew and support staff dressed in blue flying suits.

25. During the six-week Springhawk training period, the team use the clear Mediterranean skies to practise intensively – three displays a day, five days a week – as they work towards being awarded their PDA (public display authority).

26. Dye-team technicians fill the smoke pods that make the red, white and blue smoke trails used by the Red Arrows in their aerobatic displays.

27. Before a pilot straps into a Red Arrows jet, the Blues on the ground have already primed, serviced and equipped the Hawk so that it performs perfectly in the sky.

Thur 28 May

Red	Smokes	Procedures	Total
1	£5-60		£
2	£8-60		£
3	£9-20		£
4	£8-20		£
5	£10-00		£
6	£3-60		£
7	£1-40		£
8	£16-20		£
9	£21-20		£
10	£6-20		£
11	£4-40		£
APO	£2-50		£

28. An example of a 'Pigz' notepad entry, recording the fines for minor flying indiscretions against each pilot's number. This fine system is Red Arrows-specific, and intended to motivate and bond the team, while injecting a bit of humour into the intense schedule of training and displays. (Photo © Author)

29. Flt Lt Lissy Mason, the Junior Engineering Officer (JEngO), steps out of then-Red 1 Team Leader, Squadron Leader David Montenegro's Hawk, after an overseas tour. Her role as Circus 1 is to transit with the team from venue to venue and be in charge of engineering support.

30. Before climbing into the jet, the pilots will do a walk around their aircraft, making specific checks and ensuring that the physical components of the jet, including the control surfaces and hydraulic pressures, are all in perfect order.

31. Once the engine has been lit, the Hawk feels like a living, breathing thing, the needles in the cockpit are moving and you can sense all that raw potential energy about to be released. The wavy line that runs through the glass canopy is the explosive detonator cord which, in the event the ejection handle is pulled, will instantly shatter the canopy so that the pilot can safely eject from the aircraft.

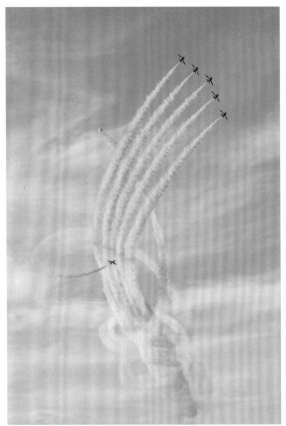

32-33. Close-formation flying and aerobatic manoeuvres are, of course, central to what the Red Arrows do, showcasing the kinds of skills and tactics that pilots have utilised since the advent of early flight.

34. The Reds regularly display at the major national air shows, such as the Royal International Air Tattoo, and at coastal events, such as the Bournemouth Air Festival (pictured) – with the latter drawing in more than a million spectators over a single weekend. (Courtesy of Getty © Finnbarr Webster/Getty Image News)

DISPLAY DIAGRAMS

FRED TO BIG VIXEN

ROLLBACKS

VIXEN BREAK

ARROW TO DIAMOND

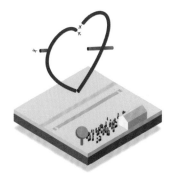

HEART AND SPEAR

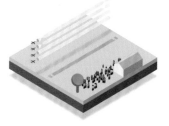

MIRROR

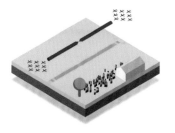

DIAMOND ROLL

BOOMERANG

PYRAMID TO SHORT DIAMOND

SHORT DIAMOND TO TYPHOON

CYCLONE

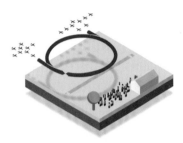

LIGHTNING TO PHOENIX

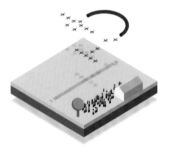

PHOENIX TO TORNADO

HAMMER BREAK

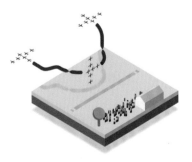

PYTHON

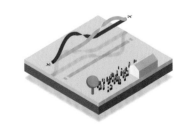

OPPOSITION BARRELL

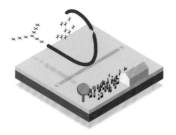

SWAN LOOP TO FRED

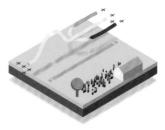

MIRROR ROLL

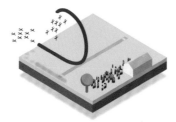

DIAMOND TO SWAN

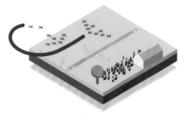

BIG VIXEN TO TORNADO

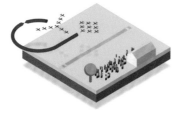

LIGHTNING LOOP

REAR SECTION BREAK

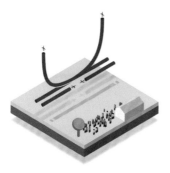

REAR SECTION PASS

GOOSE

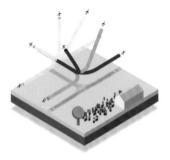

DETONATOR

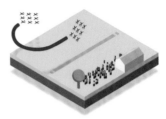

SHORT DIAMOND PRESENT

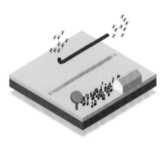

FRED PRESENT

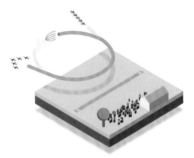

SLALOM

CROSSBOW

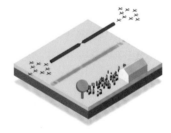

TYPHOON ROLL

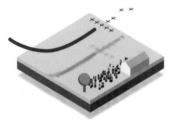

TORNADO PRESENT

TYPHOON TO LIGHTNING

even managed to convince the doctors to let him come out to Cyprus for two weeks: 'I didn't want to sit at home moping, and being with the team really helped me. Watching the Reds every day was good for my mental health.' After that Mike returned to the UK, this time to go through three weeks of intensive rehab at the Defence Medical Rehabilitation Centre at Headley Court, where the former Tornado F3 pilot was again reminded of the realities of life in the military: 'Seeing soldiers, sailors, airmen and airwomen with life-changing injuries, seeing how they were readjusting to their new lives, was incredibly motivating and mind-blowing. I couldn't mope around and, if anything, I felt like a bit of a fraud, as I'd even been able to start running again.' Although he was still recovering from his injuries, Mike did manage to get back in the red suit for some of the display season, this time in the PR team, talking to the public on the ground – a task that he really enjoyed, and which came naturally to him, as he's always had such passion for the Red Arrows.

With any incident that occurs to the Red Arrows, the team and the RAF strive relentlessly to identify what caused or contributed to the incident, to remove any potential hazard and to improve procedures, in an ongoing, never-ending effort to minimise risk. Should a team member experience a collision or accident, the last thing you want, as a pilot, is for it to define your career. Rather than saying, 'Right, you made a mistake and you're never flying with us again', the team around us and the senior chain of command help us all to move forward.

Mike's injuries in his final year as Red 6 meant that he had to have time out, but we were both asked to come back to the team: he as Red 10 in 2012, continuing in that role for six years before a season as Red 3 in 2018; and I as team leader in 2015, and subsequently as Officer Commanding of the squadron in 2020. Four years after our collision we led the biggest overseas tour the team had done for twenty-five years. My own experience of the collision with Mike, and the fatalities of Jon Egging and Sean Cunningham, undoubtedly shaped my approach to being Red 1 and the way we handle our ongoing management of risk. When Mike and I left the team together at the end of our tour in 2017, it provided a massive sense of closure for both of us. The team and the service had shown incredible faith in us, and we dared to think that we might have paid it back.

10

Flying the Flag

SHOULD YOU HAVE PASSED through Heathrow airport in recent years, you might well have been greeted in Arrivals by a series of bold images on the walls featuring iconic British people – one of them a Red Arrow, represented by then Red 9 and Executive Officer Flt Lt Joe Hourston. Alongside Joe was a Tower of London Beefeater, a London taxi driver and other familiar images of the UK, all chosen to provide a colourful and positive representation of Britain, welcoming the tens of thousands of travellers who arrive at Heathrow every day.

That the Red Arrows should be selected as one of those walled images illustrates how enmeshed the team is with the identity of Britain, both at home and across the globe. The mere sight of that red suit is enough to encapsulate all that is British – not to mention the Diamond Nine formation, or the sight and sound of nine red Hawks streaking through the sky. However, that association with the UK isn't a chance occurrence – it's the result of more than five decades of hard graft, with the team showcasing the skills of the Royal Air Force and British excellence at events around the world.

Appearances outside the UK have always formed a key part of

every Red Arrows season. These extensive tours, which are planned months (if not years) in advance, see the whole team – including a fleet of single-engine jets without any air-to-air fuelling capability – travel worldwide to perform their precision flying to gathered crowds and to attend hundreds of carefully organised ground engagements.

The Red Arrows have flown the flag for Britain all over the globe, not just figuratively but literally, too – the Hawks' tailfins have featured the red, white and blue of the Union Flag since 2015, along with the RAF lettering and roundel emblazoned on the fuselage. In striving to be one of the best aerobatic teams on the planet, the Red Arrows have been admired by aerobatic communities, air forces and spectators everywhere, and huge, excited crowds often turn out to see the team perform, with recent tours to the Asia-Pacific region and North America also reaching as many as two billion followers online.

In doing so, the team promotes the skill, capabilities and technical prowess of the RAF across the world, while also performing an essential role in defence diplomacy, reinforcing relations with key allies, boosting various British industries and fostering goodwill overseas. The positivity that surrounds a Red Arrows visit serves to deepen diplomatic and trade partnerships between the UK and other nations, and the team sometimes works in tandem with UK Defence 'hard-power' assets, as my tour to China illustrated. Since the Asia-Pacific tour of 2016, the team has also worked as a successful ally of the UK government's 'GREAT Britain' campaign, which aims to inspire people around the world to visit, do business, invest and study in the country. As a result, the Red Arrows have been involved in a variety of business, trade and educational engagements across the globe, inspiring millions of young people in the process.

ACROSS THE POND

Since their formation, the Red Arrows have combined UK displays with appearances abroad, and in fact the team's very first public display was in France, on National Air Day on 9 May 1965 at Clermont-Ferrand. The following June the team performed a display with the Lightning fighters of 111 Squadron at the Paris Air Show at Le Bourget, and further displays were held in Italy, Germany, Belgium and the Netherlands. Under team leader Ray Hanna, the Red Arrows showcased their skills in various locations in Europe, as well as in Jordan in the Middle East. There the team put on a display for King Hussein, himself a pilot and a long-time supporter of the Red Arrows. Once they had established themselves as a worthy international display team, the Reds continued to travel abroad and became a star attraction at the many air shows held in Europe, where they would display alongside other notable aerobatic formation teams, including the nine-ship Patrouille de France of the French Air and Space Force and the ten-ship Frecce Tricolori of the Italian Air Force.

In 1972 the Red Arrows made the first of many tours outside Europe, crossing the Atlantic for a promotional tour of the United States and Canada. Squadron Leader Ian Dick would lead the formation on a tour known as 'Operation Longbow', which would push the single-engine Gnat to the extreme. With its under-wing fuel tanks fitted back on (they were removed for display flying), the Gnat had a maximum range of 500 miles. To make it across the Atlantic the team had to break the journey into three legs, flying from Stornoway on the Isle of Lewis to Keflavik in Iceland and then on to Sondrestrom in Greenland, before reaching Baffin Island in Canada. On each leg of the crossing – the longest being 450 miles – the team reached a point of 'no return', where limited fuel meant they couldn't return to the place they had taken off from, the team's only option being to push on to the one and only airfield ahead of them where they could land.

Without any navigation equipment on the Gnats, the team was reliant for logistical support on two Hercules aircraft, which between them carried almost a complete Gnat's worth of spare parts, as well as dozens of tyres and various other necessary items, not to mention a whole team of support engineers. Each stop required the ground-crew engineers to service the aircraft in bitter conditions, sometimes at twenty degrees below freezing, with biting winds adding to the chill factor.

Team leader Ian Dick recalls that the transit to North America went more or less according to plan, but the route back resulted in the 'skoshiest' moment ('skosh' being old-school jargon for poor visibility conditions) of his flying career: 'We were due to land at Keflavik in Iceland – there was no other airfield where we could land, and we had limited aid to get us down, so it had to be a ground-control approach, in which air-traffic controllers guide a pilot down to the runway. We couldn't talk to Keflavik, but our support aircraft could, so the system was that one Hercules would go ahead to the airfield, check the weather was all right and then radio back to the Hercules escorting us, to let us know the weather was okay and it was fine for us to proceed. What we didn't know was that the lead Hercules had just circled the airfield and hadn't done an approach, and when we actually got there, the weather was below limits – both clouds and visibility were below the usual minimum requirement for an approach.

'We were flying in two groups of five: I was leading the front five. As the rear five had less fuel than we did, I told them to go ahead and do the approach while we circled above thick clouds that seemed to cover the whole of the North Atlantic. Looking out of my window, I suddenly saw a hole in the cloud the size of a football pitch, through which I could make out the sea. Remembering what Ray Hanna had once said to me – "Ian, the sight of the ground is worth a thousand ground-control approaches" – I radioed to the front five to follow me, air brakes out, rolling over, and we descended from about twenty-five thousand feet through this gap in the cloud, which was about twenty-five miles west of Keflavik. I have no idea how I found it but, suffice to say, a few minutes later I ran in over the lights of the airfield, right on the centre line. The other section had to use a ground-control approach, each turning individually and landing in a clatter of bits.'

The tour itself was a great success, with the team travelling to the Canadian Air Force bases of Goose Bay in Labrador and

Trenton in Ontario – where the Red Arrows flew over Niagara Falls – before heading south to various shows in the United States. The team was met with great enthusiasm at the Reading Air Show in Pennsylvania and at the Transportation Exposition at Washington DC, which drew in daily crowds of more than half a million people. The display there included a fly-past in Concorde formation, before changing into Feathered Arrow for a loop. Still the only team in the world to roll five in Line Abreast, the Red Arrows impressed the crowd with a wine-glass roll – the synchro pair performing a variety of high-speed crosses – before all nine aircraft joined together again for a loop and a final bomb-burst to mark the end of the show. In showcasing the capabilities of the RAF, the tour was deemed a great success, further cementing the reputation of the Red Arrows and the UK overseas.

The arrival of the Hawk in 1979–80 – a far more fuel-efficient jet than the Gnat – enabled the Red Arrows to travel further distances. To make the most of this new aircraft, the team embarked on its first Middle Eastern tour with the Hawk in April 1981 under Squadron Leader Brian Hoskins. The Reds' first display was set for Abu Dhabi in the United Arab Emirates, with the team stopping to refuel in Saudi Arabia along the way. Diplomatic discussions led to delays over where this stop might be, with agreement finally being reached on Tabuk in the north-west of the country. However, the team was under local orders to refuel and leave the airfield before dark. When they got back into the Hawks, ready to head to Abu Dhabi, it soon became apparent that five of the aircraft wouldn't start, at which point the team's trusty engineers set to work. Suspecting that the fault had something to do with the strong winds and sandstorms in the area, the engineers soon discovered that sand had got into a butterfly valve under the starboard wing of each aircraft, which prevented it from opening correctly during the engine starting cycle.

'The Saudi authorities allowed the team to stay overnight,' relates Brian Hoskins, 'but we would need to leave before dawn the next day. Thanks to the engineers, we were able to do that, and we took off into a stunning sunrise over Saudi Arabia, making the display in Abu Dhabi.' The team then flew low-level to the brand-new airport in the city of Dubai to display there, and afterwards in Sharjah. The next day it displayed in Bahrain, with both Kuwaitis and Saudi Arabians among the spectators. The Red Arrows then had a planned stop in Jordan, where they flew a display for the Jordanian Royal Family. Hoskins also took sixteen-year-old Prince Faisal, son of King Hussein, in his back seat, performing a few loops and rolls.

Summer 1983 saw the team return to Canada and North America, this time under the leadership of Squadron Leader John Blackwell. Still requiring fuelling stops along the way, the team flew from RAF Kinloss in Scotland to Canada via Reykjavik in Iceland, then spent their first night at Sondrestrom in Greenland. When flying over or landing in Greenland, pilots are often struck by the dramatic scenery of the country, as Blackwell related in a 1983 documentary about the trip: 'Sondrestrom was a fascinating experience. The airfield itself is very sparse and sits at the end of a hundred-mile-long fjord . . . I had this overriding impression of the ice cap of Greenland and this beautiful fjord still frozen over in early May. Immediately to the left of the airfield is a thirteen-hundred-foot-tall mountain . . . it really is the most spectacular airfield.'

The team went on to refuel in the now-familiar Goose Bay in Canada at the end of a journey over the Atlantic that still required just as much nerve and focus as Ian Dick's team had shown back in 1972: 'There were a lot of calculations done, believe you me, and when you look down and see ice floes beneath you, and you have only one engine and not a hell of a lot of fuel, it tends to concentrate the mind beautifully.'

Four days' rest in Trenton, Ontario were followed by thirteen displays, including at Andrews Air Force Base just outside Washington, where the team displayed alongside the US Air Force team, the Thunderbirds, in their F-16s in front of an enormous crowd of around half a million people. Then it was on to South Carolina, Florida, Texas, Georgia and finally Annapolis in Maryland, where the Red Arrows flew with the US Navy aerobatic team, the Blue Angels.

Never ones to settle for long, three years later, in 1986, the Red Arrows embarked upon their longest-ranging tour, displaying in Europe, then in Egypt and the Middle East, before heading to Asia via India, Malaysia, Thailand, Singapore and Indonesia. With the end of the Cold War, in 1990 the team toured in Russia, Ukraine and Hungary for the very first time and met a warm reception. The Red Arrows also displayed at the Czech and Slovak International Air Fest (having never performed in the former Czechoslovakia before) alongside other national teams, including the Slovakian display team, Biele Albatrosy, the Spanish Patrulla Águila and the French and Italian display teams.

AROUND THE WORLD

In 1994 new team leader John Rands received a letter from the leader of the South African aerobatic team, the Silver Falcons. It was inviting the Red Arrows to celebrate the seventy-fifth anniversary of its air force, which came into existence at roughly the same time as the RAF, so the two share quite a bit of history. Apartheid was coming to an end in South Africa and so, as John puts it: 'I thought it would be a great adventure to go, but considered it highly unlikely, as the proposed date of October 1995 was outside the normal display season.' In the autumn and winter months the engineers usually had most of the team's aircraft in for deep servicing, which often left the Reds with about four or five serviceable aircraft at a time. As a result, the team generally had to borrow a couple of 'rentals' from RAF Valley, which they endeavoured to return in a better state than they were supplied. There simply weren't enough aircraft to facilitate the South African trip, until John managed to persuade the senior chain of command to provide one extra aircraft, bringing the total pool up from eleven to twelve Hawks and enabling the tour to go ahead.

With the tour funded by twelve to thirteen British companies, including British Aerospace, Rolls-Royce and the ejection-seat manufacturer Martin-Baker, the team set off for South Africa at the end of the 1995 season. It headed through Europe to display in Greece and Turkey, before flying on to Qatar, Bahrain and Oman, where in a very hot Muscat, to their surprise, they came across HRH Prince Charles, who was with the Omani Royal Family, wearing (as John puts it) 'a wonderfully crumpled cotton suit that only a royal could carry off and look distinguished'. After performing a display for the royals – Red 7 landing almost on fumes as they were so low on fuel – the Blues serviced the aircraft and the Red Arrows all headed off that evening. They needed to keep moving if they stood any chance of making it to Waterkloof in South Africa in under forty-eight hours, in time to attend the South African Air Force's seventy-fifth birthday celebrations.

So the next day the team pressed on to Jeddah, then Addis Ababa in Ethiopia, which, John said, was 'completely outside my comfort zone, due to the unique conditions. The airfield was nearly eight thousand feet above sea level, the weather was very different to anything we were used to, and we had little in the way of navigation aids or air-traffic control. I also remember that in coming to a halt on the side of this long runway and waiting for the rest of the guys to land, I saw a family appear out of the long grass, who waved at me and I waved back. When we later took off, they were sitting on the side of the runway.'

The team arrived at Jomo Kenyatta international airport in Nairobi that evening in torrential rain and cloud, with the eleven aircraft travelling in two sections (five at the front, six at the back) – and with visibility so bad that at times during the transit John Rands, leading the front five, couldn't see the other aircraft on his wing. The weather over the African continent usually proved challenging, as a result of the Intertropical Convergence Zone. This is a band of low pressure that normally sits near the equator and causes heavy rain and huge thunderstorms. Some of the storms go up as high as 50,000 feet, which the team couldn't fly over or, with limited fuel, around.

After Nairobi, the team hopped from Lilongwe in Malawi to Waterkloof in South Africa, arriving at 2 p.m. and then flying a display later that afternoon to an enthusiastic and friendly crowd of more than half a million people. In under forty-eight hours the Red Arrows had flown 4,200 miles and, with a display at the beginning and end of the journey, had packed in as much as they could on an overseas tour, as we still do today. On top of this demanding schedule, the team had to cope with 'hot and high' conditions at Waterkloof – temperatures well above thirty degrees Celsius, and an altitude of more than 5,000 feet above sea level – which reduces the available thrust from the engine, making it very challenging to fly looping manoeuvres. 'It took longer to get airborne, longer to climb to display height, and maintaining the right display speeds was a struggle,' remembered John Rands. 'As leader, I was using far higher thrust settings than normal, which meant the wingmen had reduced excess thrust margins to maintain position. I quickly learned just how much thrust I could steal and keep everybody aboard. The shows in Waterkloof were about four minutes longer than the shows flown at sea level, as I flew slightly wider patterns to conserve energy. The boys did a magnificent job to display so well in such conditions.'

After a week in Johannesburg, the team hopped down to Bloemfontein and Cape Town, where they displayed at the waterfront with Table Mountain looming in the background, and then flew to Durban along 'the Garden Route' of the south-eastern coast of South Africa where, flying at low level, they could enjoy the beautiful views of bays, lagoons and waterfalls cascading into the sea – the 'trip of a lifetime', as Rands describes it. From South Africa, the team worked its way up to Zimbabwe, landing in Harare, where it put on a display in front of a large and enthusiastic crowd. After performing a fly-past over Victoria Falls the Red Arrows headed home, covering 800–900 miles on each leg, with three legs a day. A trusty Hercules transport aeroplane shadowed them along their demanding journey, although the team never had to call on its support, as the Circus engineers managed to keep all the Hawks serviceable the whole way home. The team returned to RAF Scampton for few months of winter training, before flying to Malaysia to appear at an international air show on Langkawi Island in early December.

After Christmas the Reds flew from Malaysia in stages to Australia. Arriving first in Darwin, the team then flew down to Adelaide via Alice Springs and on to Sydney for the final display. The Red Arrows flew over Sydney Harbour on 26 January 1996, Australia Day, with a record 1.2 million people watching as they flew over the iconic sights of Sydney Opera House and the Sydney Harbour Bridge. They then made the long transit home, attending a Singapore air show in February and arriving back to a snowy RAF Cranwell in late February. The world tour of 1995–6 had seen the team travel 52,000 miles – the equivalent of twice around the world at the equator – and a staggering 1,100 sorties. It is a tour that is still talked about to this day, and a monumental achievement for the Royal Air Force Aerobatic Team.

STORMY WEATHER

When I returned to the team as Red 1 in 2015, plans had already been set in motion for a major Asia-Pacific and Middle East tour the following year, one that would cover 20,000 miles and seventeen countries. This would be my first long-haul/global tour with the team, not to mention leading the formation to China for the very first time. While our support team was putting the usual extensive preparations into place, working closely with our governmental allies, there was a sense of launching into the unknown, when it came to China, and I was mindful of the logistical and diplomatic challenges that potentially lay ahead.

This was the 'golden era' of UK Prime Minister David Cameron and China's President Xi Jinping, with greater trade and investment deals between the two countries, and this tour formed a central part of overall UK engagement in China, as well as of the Department for International Trade's 'GREAT Britain' campaign.

In effect, the Red Arrows were there to foster even better relations with China, to push the prosperity agenda for the UK and to open new doors for trade and wider British interests. The team had a strategic and diplomatic role to play, the sensitivities of which were borne out when the team took off from Da Nang in Vietnam to cross the Chinese border. On the same day RAF Typhoons were also crossing the South China Sea, heading to Japan and South Korea to conduct a multinational training exercise. So just as twelve red RAF aircraft were crossing into China, in a 'soft-power' bid to support the UK/Chinese relationship, 'hard power' in the form of RAF front-line aircraft were heading towards strategic partners in the area to train with them as part of the Exercise Eastern Venture. The two operations made an important statement: while the UK wanted to trade and promote its prosperity agenda in China, and demonstrate British excellence at its best, it was still very much a global power with strategic partners all around the world.

As with all our long-haul tours, we don't simply pack up our jets and ship them where we need them – we have to physically get them there by air. To carry spares, equipment, engineers and personnel, many of whom had never flown further than the Middle East, we also needed a couple of Hercules aircraft – the workhorse of the RAF transport fleet, which have accompanied the Red Arrows on overseas tours for many years and is now retired from service. To make the 9,000-nautical-mile journey from the UK to China, we needed to refuel no fewer than twenty-two times on the way there, and twenty-four or twenty-five times on the way back – because of the way the Earth's rotation influences wind patterns (namely, the high-altitude jet streams), flying from east to west always takes that bit longer. Many of the refuelling stopovers also involved displays and ground engagements, with the task as always being to squeeze in as much as we could on the journey. The plan was to engage on the way east primarily in Jordan, India, Pakistan, Bangladesh, Singapore and Malaysia, and on the way home to do another round of engagements in India, Bahrain, Oman, the UAE and Kuwait.

From day one we experienced operational challenges, some of which really did come out of the blue. For every tour we build in rest days with transit days, always with contingencies for unknown issues along the way. We can't build in too many of these days without flying or engagements, but need to strike a balance between fatigue management and maintaining a high-tempo and visible campaign. Launching from Scampton on a warm October day, we staged through Europe, using Cyprus as an overnight refuelling stop, before flying over Israel into Jordan and then launching again for two refuelling stops in Saudi Arabia. However, on the day of departure from Cyprus, former Israeli Prime Minister, Shimon Peres, died and Israel shut its airspace for seventy-two hours, leaving us without anywhere to go; not a great start, as we missed our first engagement in Amman, Jordan.

When we were finally able to leave, we modified the plan to play catch-up and managed to get from Cyprus to the east of Oman in one pretty gruelling day, involving a dawn take-off, three refuels across Saudi Arabia and a dusk landing in Oman, by which point we were pretty done in and still acclimatising to the searing temperatures of the Middle East. Our next big engagement was in New Delhi in India, where we were to attend an ambassador-level engagement at the British High Commission, which formed part of the government's 'GREAT Britain' campaign.

To get to Delhi from Oman, we needed to refuel in Karachi in Pakistan, then again in Jaipur, reaching Delhi in one day. We took off very early and were faced with poorer weather than was forecast on arrival into Masroor air force base, Karachi: the city was enveloped in smog and haze. It was just about safe for us to approach the airfield, but it nonetheless proved incredibly challenging, diving down into a gap, similar to the way Ian Dick had to do when he led the team into Keflavik, Iceland, back in 1972. Having phoned our Pakistani compatriots that morning, confirming exactly when and how we'd be arriving into Masroor, we were surprised to discover, on nearing the air base to the west of Karachi, that the Pakistani Air Force was running a large-scale military exercise that day. As I attempted to lead twelve aircraft into an unfamiliar airport – a feat in itself – there were multiple Pakistani fighter jets everywhere around us as we were coming in to land. As if this wasn't bad enough, there were also huge crane-like birds everywhere, which feed on the wastelands around the city and were flying directly under our wings as we approached the airport. Not only did we have to manoeuvre around the fighter jets, but we were also busy dodging what the pilots later referred to as 'pterodactyls' trying to take our wings off. When we all managed to touch down safely, I was beyond relieved, although I was very aware that we'd completed only the first of three legs that we were doing that day.

Before heading off for the second leg of the day, the Circus engineers needed to turn the aircraft round, refuel them and do any light maintenance, all in an enervating thirty- to forty-degree Celsius heat. With that done, we headed off to Jaipur in India. I was aware that there were some thunderstorms brewing, but I decided it was safe to go, thinking that we should be able to manoeuvre around any bad weather. Once we got airborne, however, the link with Karachi air-traffic control went from bad to worse, and it was almost as if we were talking to someone sitting in a tin cabin. We were getting increasingly broken messages from Karachi radar control until, at about sixty miles to go to the Indian border, we lost all communications with air-traffic control. I could now also see, sitting right on our nose, the threatening dark mass of a huge thunderstorm.

I had two options: turn back to Karachi or navigate around the thunderstorm, a decision I needed to make knowing that we had no radar, effectively no air-traffic control and were rapidly approaching the border between two countries that were at conflict over Kashmir. Leading twelve military aircraft across the border and unable to communicate with anyone, I knew there was a chance that we might be intercepted by an Indian fighter plane. Weighing up the fairly dire options, I decided it was safer to press on; and, as we did so, one of the team pilots searched every Indian air-traffic control frequency that he could before eventually, to our relief, making contact with one. It asked us for our position, which we gave, and then there was a very long pause – enough for me to think: 'Right, we could be in a bit of trouble here.' Finally air-traffic control radioed: 'You have bad weather. Bad weather. You cannot route north.'

The sight before me confirmed the worst. What had looked like a single cell of thunderstorms ahead now towered menacingly to the very top of the atmosphere and seemed to reach at least 200 nautical miles east. There was no way we were going to be able to get through it. As we didn't have enough fuel to return to Pakistan, our only option was to find an airfield to which we could divert and land. This turned out to be a small regional airport at Ahmedabad in Gujarat and – not long after getting radio contact from a foreign squadron it had never heard of – Ahmedabad airport had slightly under twenty-one minutes to prepare for a twelve-aircraft landing.

Meanwhile, we battled through the bad weather, manoeuvring around the enormous, rumbling cumulonimbus clouds. This involved picking a blue gap and leading the formation through it, which feels a little like flying through a vast canyon – the problem being that you're never quite sure what lies in store at the end of it. On one of these final repositions I clipped the outer edges of a huge cloud and we were immediately thrown at least ten to twenty feet in every direction. Ice quickly formed on our windscreens and I lost all front visibility. I could just about see Reds 2 and 3 managing to hold onto my wing, but I'd never before (and never since) seen them move vertically up and down quite so rapidly. After about five to ten seconds we were out the other side, but for a few heart-pounding moments my fear was that I would lose the rest of the team or that the engines would ice up at the front – a disastrous outcome for anyone flying a single-engine aircraft.

As we neared Ahmedabad airport we separated the formation into four sections, each made up three aircraft, which is always the safest approach for landing. This in itself needs to be done properly, as the worst scenario would be for a section to collide with another one in cloud. Once all four sections were safely spaced out from one another, we set up for the instrument approach to the runway, although the poor weather made for a pretty dicey landing. However, all twelve of us made it down safely and, as we taxied in, I could see great pools of water on the ground from the monsoon of rain that had just passed through. For the second time that day the relief of finding somewhere to land and getting down safely was immense, and we had to hand it to the Indian air-traffic control who, considering that we pretty much came out of nowhere, handled us brilliantly as we pitched up at a small airport that probably saw no more than about twenty aircraft landings in a day.

We ended up staying in Ahmedabad for three nights and, with accommodation hurriedly sorted, the team travelled through the brilliant chaos of the city in an assortment of local taxis and tuk-tuks. Slumping into a chair in my room, I replayed the day's events over and over again in my mind, questioning whether I had made the correct decisions in what had been a very tricky twenty-four hours. Every team leader is faced with hugely challenging days during their tenure, dealing with a mix of operational issues, political pressures, personnel issues and the need to make vital decisions that, if made incorrectly, could have devastating outcomes. This was definitely one of those days for me. Over the next two days, on looking at the weather forecast and the storm activity ahead, I had to make the decision that ultimately it wasn't safe for us to progress. The delay in Ahmedabad meant the pilots missed an embassy event in Delhi, but greater forces were at work. Thankfully our advance party in Delhi carried out the engagement without us, in the spirit of true Red Arrows teamwork.

Eventually we did make it to the air show in Delhi, meeting up with our Indian display-team counterparts, who also flew the Hawk – and some of whom were students I'd taught a few years before. The smog there resulted in terrible visibility again, but we managed to display safely, and thereafter our transit east-bound was much smoother, although finding the right parking spot at

Calcutta international airport proved more difficult than the flight over there.

After Calcutta, our next refuelling point in Dhaka, Bangladesh unearthed some tricky operational issues en route to China. We had been asked to conduct a fly-past at Hazrat Shahjalal international airport at Dhaka, where the visibility was really poor, just three miles or so, which meant that finding the airfield was a challenge in itself. Leading the team along a taxiway surrounded by tropical plantations was something of an adventure, although when we saw an old Russian MiG-21 fighter painted in Bangladesh markings I knew I'd found the right spot. We shut down the engines and jumped out of the aircraft, whereupon the Bangladeshi Air Force gave us the warmest and kindest welcome possible. Handshakes were followed by a discussion of our operational experience and a lavish lunch, although all the while my mind was thinking about the rest of the day, knowing that the crossing of Myanmar lay ahead of us, before we could land safely in northern Thailand.

Fully aware of the political instabilities in Myanmar at the time, we had been cautioned that any emergency diversion in the country would incur some tricky diplomatic issues for Her Majesty's Government. The weather provided another layer of complication: thunderstorms were building quickly and I was keen to leave Dhaka as diplomatically and as quickly as possible before what would turn out to be one of our longest and most stressful transit sorties – at least for me, anyway. Once airborne and over Myanmar, we observed some fires on the ground and I couldn't help but wonder what might be going on beneath us, but after an uneventful hour in the air we made it to the border with Thailand, where we had to carefully re-route around some thunderstorms that had bubbled up and were blocking our path. We eventually made it through and landed at Chiang Mai just as the sun was setting. Stepping out of the aircraft, I felt a huge sense of relief and thought to myself, 'Great, we've got through the most challenging part of the transit.'

That relief was short-lived, however, when not long after landing, and as we all sat huddled in a small room, a phone was brought to me. On the other end was the British Defence Attaché of Thailand, who informed me that the King of Thailand had died that afternoon. He went on to say that the Thai people were hearing about it that day and he had no idea when we would be able to fly out of the country. Yet again we were stymied by events that few people could have foreseen – although the unexpected is often a feature of long-haul tours with the Red Arrows – and assumed we would be held up for a while. The Thai authorities, however, were incredible and, at a time of major national mourning, helped us to get through Thailand. With their assistance we pressed on to Singapore and then to Malaysia, where we put on a display in downtown Kuala Lumpur, performing a fly-past of the Petronas Twin Towers, before heading north into Vietnam, which in itself was enormously challenging.

MAKING HISTORY

By this point the team was getting very tired – I noticed they were napping whenever they could, on minibuses or wherever we were. Some were visibly losing weight – I think I lost about thirteen pounds in ten days – and food poisoning had become an issue along the way, so managing the cumulative touring fatigue was at the forefront of my thoughts.

Flying over the vast jungles of Vietnam, with the magnificent green canopy of trees and the humidity rising, was an incredible but forbidding sight from a single-engine Hawk. Our survival packs, tucked under our ejection seats, were now equipped for emergency needs in a jungle, with knives, insect repellent and the like. Absolutely no one wants to end up in a jungle survival scenario, and this overriding thought propelled the team towards the coastal city of Da Nang. On landing, tired as we all were, a certain amount of adrenaline had kicked in, as we knew this was our last stopping point before heading across the Chinese border – a historic first for the Red Arrows – and we were only hours away from making it there.

Our destination airport in China was Zhuhai, just west of Hong Kong, on the mainland, bordering Macau. Chinese regulations separate aircraft using metres, not feet, the latter being the predominant system used around the world. So to counter the fact that all our altimeters read only in feet, we stuck paper conversion tables to the front of our glass canopies to ensure that we flew at exactly the right clearance height. As we approached the border and made contact with Chinese air-traffic control, I let my mind take in the situation. I couldn't believe that I had been given the responsibility to lead us into China for the first time in the team's history.

It was exhilarating, but that feeling soon passed as air-traffic control ordered us to descend and re-route, and mild panic began to take over from what had been pride and elation on making it into Chinese airspace. The team quickly established that if we were to comply with its order, our jets would run out of fuel before we could make it to Zhuhai. The air-traffic controller sternly repeated his order, to which the Red Arrows responded, 'Negative, we cannot comply.' The controller did not reply to us, so I sat nervously thinking that we were potentially about to start a diplomatic crisis within our first two minutes of being in China. I told the team that we would continue with our original flight plan and asked them to keep their eyes open – I really didn't know what might happen next.

We safely tracked east and, with communications among the team kept to a minimum, the formation flew in eerie silence. After an hour the controller piped up again and told us to contact Zhuhai approach frequency. We tentatively radioed our call sign, our height and our present position. A very friendly and seemingly excited controller welcomed us to China and to Zhuhai, and our descent into Zhuhai went as planned. With mainland China filling the view through my canopy, I ordered the team into close fly-past formation and to smoke red, in honour of our hosts. In one of the most memorable 'Smoke on, go!' calls of my tenure, I

watched Zhuhai whistle past beneath us as we performed our signature airfield arrival.

We had in fact made it to China bang on time. We'd had to
use up all the contingency time and were right up to the wire, but
all twelve aircraft had made it and no one had been left behind.
As if that wasn't enough cause for celebration, on landing we
were greeted with the warmest, most welcoming reception that
we'd probably received anywhere in the world. We had no idea
what to expect, but were overwhelmed by the number of aviation
enthusiasts in China – people who had followed the Red Arrows
all their lives, and who knew all the team pilots and everything
about us. For me, our time in China really brought home how
far-reaching the enthusiasm for aviation is, and how it can bring
people together, whatever the political situation.

The potential reach of the visit was also brought home almost
as soon as we stepped out of our Hawks: China Central Television
had filmed us arriving in Zhuhai and had requested interviews
with me and the other pilots, which they then packaged up and
put on their breakfast TV over three days. We discovered later that
each of these five-minute segments had a viewership of more than
half a billion people – a staggering 1.5 billion viewers in total. In
Zhuhai we were also asked to attend several other media engagements and were met with real excitement and reverence, the like
of which I can only imagine film stars are used to; it was certainly
beyond anything we'd ever experienced before.

Our arrival in China was followed by five days out of the
cockpit – much needed, after the gruelling journey – although we
were all still very much at work, attending a number of ground
engagements that formed part of the UK's 'GREAT Britain' campaign. Within our first twenty-four hours in China we visited
universities, schools and various industries and met around 5,000
people. I was selected to attend an event at the embassy in Beijing
and, not long afterwards, found myself back in the red suit, this
time for an interview with *The Sunday Times* on the Great Wall
of China.

In Beijing we also attended a STEM engagement organised by the British Council at the Beijing Institute of Technology. Many of the students who were there were filming us on their phones. I remember our Group Captain leaning across to one of the students to ask, 'Are you recording this for yourself?' 'No,' the young man answered, 'I run a YouTube channel. Two hundred and fifty thousand people are watching live, right now.' He was just one student out of hundreds there – the scale was out of this world, and this was before we'd even displayed in the country.

After a few days we were back in the Hawks and performed a total of nine full shows at the China International Aviation & Aerospace Exhibition in Zhuhai, where we met our Chinese display-team counterparts. After a brief stop in Hong Kong for some school visits and industry engagements, we travelled back to the jets in Zhuhai and began the journey home, which also came with multiple pit stops along the way to refuel. We ensured that we made the most of every moment, fitting in more displays in India and a huge school engagement in Hyderabad, before flying back through the Middle East. By this stage, twenty-one coun- tries in and having travelled pretty much non-stop for nine and half weeks, it felt like a never-ending tour – but we made it home, flying over the green fields of Lincolnshire into RAF Scampton on a cold, early-December day.

Within the team we felt that the tour had achieved everything it had set out to do and more, and we were blown away by the enthusiasm and huge numbers who had engaged with our visit. The embassy in Beijing had helpfully tracked and analysed the impact of the visit, concluding that in the six months after our arrival about £7–8 billion of business activity had developed, not only as a result of the Red Arrows visit, but as part of the whole UK engagement plan around that time, in which we had played a significant part. While it has always been acknowledged that the Red Arrows have a key ambassadorial and goodwill role overseas, that kind of financial impact put into sharper focus what a team such as ours can do to help further governmental objectives. On a military level, the visit also helped to forge closer ties between the UK and Chinese military, with a number of Chinese three-star officers visiting RAF Scampton not long after the tour.

LONG HAUL

Over the following eighteen months planning had already begun for the Reds' next major tour of Canada and the United States. Covering more than twenty-five cities over eleven weeks, it would prove to be the Red Arrows' biggest-ever tour in North America – one that continued to support the 'GREAT Britain' campaign, championing a range of UK interests across trade, business, education and defence, while bolstering the UK's relationship with our long-time allies. Squadron Leader Martin Pert, who had taken over from me as Red 1 for the 2018 season, was thrilled to be leading the team on the tour, and particularly relished the prospect of taking twelve single-engine jets across the Atlantic.

Today the team faces the same challenges as those encountered by the likes of Ian Dick and Dickie Duckett back in the 1970s: flying single-engine jets at their maximum fuel range, at the whim of Atlantic weather. 'When I first began flying at an aerodrome, flying little planes and "puddle-jumpers"', recalls Martin Pert, 'the dream for most was to fly across the Atlantic, land in the US and say you've flown there yourself. Flying something like the Hawk is very different to flying in an airline or front-line jet, where you can air-to-air fuel along the way. It really is an adventurous feeling and, on a tour like that, getting there and getting back are the biggest challenges.'

Having completed a shorter, yet no less packed domestic season, the team embarked on the usual three-leg trip across the Atlantic in early August, flying from Lossiemouth in Scotland to Keflavik in Iceland, Narsarsuaq in Greenland and then on to Goose Bay in Newfoundland, Canada. Along the way the pilots encountered challenging weather, in the form of thunderstorms and heavy rain, as well as technical issues: one aircraft's nose wheel had stopped functioning properly, meaning that it had to stay behind in Iceland. The team had no choice but to press on as an eleven-ship formation, with the twelfth aircraft and pilot re-joining them later in the trip. Landing in Greenland, this time at the isolated airfield of Narsarsuaq, was as much an awesome experience for Pert as it was for previous leaders of the Red Arrows, as Martin recalls: 'Flying down a fjord in Greenland with ten-thousand-foot ice-caps either side of you, icebergs in the water, there's a huge sense of relief when you can see the runway, as that really is your only option for landing.'

The 2,658-nautical-mile journey across the Atlantic took the team three days, with the jets arriving in Halifax, Nova Scotia to the minute, ready for their first public performance on 11 August 2019: an enhanced fly-past over its waterfront. Thereafter the team performed at the Gatineau-Ottawa Air Show, at the Air and Water Show in Chicago, in Boston, Atlantic City and then New York City, where the Reds performed in a nineteen-aircraft flight along the Hudson River and around the Statue of Liberty, painting the Big Apple's skies with their red, white and blue smoke. The Red Arrows were led by the F-16s of the US Air Force Thunderbirds and had two F-35 Lightning IIs and a pair of F-22 Raptors tucked in behind them, which made the flight even more exhilarating, both for the team pilots and the spectators below. Red 10, Adam Collins, with a photographer in his back seat, managed to capture some extraordinary images in the process, describing the event as 'the best hour I've ever spent in an aircraft' – which is quite something, coming from an experienced front-line pilot with some twenty-two years' service in the RAF.

In touring the United States it does become apparent that the Red Arrows display is quite different in style from those of its US counterparts. The US Air Force Thunderbirds in their six F-16s, and the US Navy demonstration squadron, the Blue Angels, who fly six Boeing F/A-18 Hornets, both perform breathtakingly tight formations and aerobatics accompanied by loud music played to the crowds. The Red Arrows, flying in a larger formation and in simpler fast-jet aircraft, have always relied solely on commentary from Red 10 and the use of coloured smoke. It's a more understated style, and the smaller ground crew needed makes for a nimble and flexible operation, so that the Red Arrows are able to perform at multiple events over the course of a single day.

After a string of exhilarating performances and public events in Washington DC, the team flew back to Canada, performing a fly-past in Battle formation over Niagara Falls, before heading to Toronto. During a practice for the Canadian International Air Show, Canadian astronaut Colonel Chris Hadfield flew with Red 1, Martin Pert, during a joint fly-past with the Canadian Forces Snowbirds over Toronto's shoreline. The former military pilot turned astronaut seemed to enjoy his flight with the Reds, remarking afterwards, 'That was a joy, hilarious and fun. So tight, crisp and professional' – resulting in another special moment for Martin and the team. The Red Arrows also went on to celebrate Chris Hadfield's sixtieth birthday by attending and watching a Blue Jays baseball game after they were presented to the crowds.

Thereafter the team stopped in Ohio, then flew on to a very hot St Louis in Missouri, before heading to Houston and Fort Worth in Texas. From there the Reds did a short transit to Denver, where upon landing they could hear air escaping from seven of the team's tyres as they promptly deflated. This was a problem they'd never had before. The tyres had expanded with the high altitude of Denver, nicknamed the 'Mile-High City' because it's a mile above sea level, and during the short flight from Texas the tyres hadn't had enough time to cool down before heating up again on landing, which caused the overheated tyres to deflate. Fortunately the ever-prepared Blues had eight spare tyres, which were swiftly fitted.

The team subsequently pushed on to Portland, Seattle and then Vancouver in British Columbia, Canada, before performing at the Miramar Air Show in San Diego, California – the strike-fighter training base made famous in the movie *Top Gun*. Just as much of the crowd was excited to see the Red Arrows, so many in the team were thrilled to be performing alongside aircraft such as Oscar,

Cobra and Viper helicopters and F-35 Lightnings at a place that inspired many members of the team to pursue their love of flying. Red 3, Squadron Leader Mike Bowden, was thrilled to visit Miramar. On what was his fifth and final tour with the Red Arrows, he explained, 'Growing up, I was fascinated and inspired by two things: one was watching *Top Gun* and this place, and the other was the Red Arrows.' As it's Red 3's job to gee-up the team as they approach a fly-past or a display, Mike's chosen words as they neared Miramar had to be, of course, 'I feel the need . . . the need for speed!'

The team went on to fly over San Francisco's Golden Gate Bridge, the Grand Canyon in Arizona and the Hollywood sign in Los Angeles. Finally, after eleven weeks away, having appeared at more than 120 events, performed two displays, thirty fly-pasts, flown 22,000 miles and reached some two billion people across a variety of social-media platforms – including a ten-episode BBC podcast and a four-part television documentary series – the Reds were ready to come home. The 108-strong team of pilots, engineers and support staff, twelve Hawks and one Atlas A400M RAF transport carrier were packed up and made ready for the journey back over the cold waters of the Atlantic.

For many in the team, seeing the spires of Lincoln Cathedral – the famous run-in sight for RAF Scampton – is a welcome moment, and everyone was looking forward to seeing their family and friends after a long stint away from home. But with new pilots joining, a fresh season to plan for and future Red Arrows overseas tours on the horizon, new adventures are always calling!

11

Marking the Occasion

SINCE 1965 THE Red Arrows have helped to mark and cele-brate a huge range of national events in the UK, from RAF and military anniversaries to sporting events, the unveiling of national landmarks and royal commemorations. A fly-over by the Red Arrows turns an occasion into something very special, while also providing a degree of constancy and assurance that, despite what-ever else is going on in the world, the Reds and the Royal Air Force are still there for us. A welcome sight in our skies, the Red Arrows also reach millions through their appearances on television, in print and across social media; younger fans can even experience their own white-knuckle Red Arrows Sky Force ride at Blackpool Pleasure Beach.

The Red Arrows alone make for a spectacular display, but two or three times a year the team joins other military or commer-cial aircraft in mixed formations, often to help celebrate a British aviation venture, promote engineering excellence or mark a sig-nificant anniversary. Put nine red Hawks on the wing of another aircraft and you'll have a mass-media event, and if it's an aircraft that's recently come into service or is about to go into retirement, a

very poignant one for everyone involved. In marking the achieve-
ments of British aviation, certain heavy aircraft have regularly
flown as 'Red 11' with the team, including a Second World War
Avro Lancaster, a Vulcan V-bomber and the iconic supersonic
airliner, Concorde.

Back in 1973 Concorde was still undergoing test flights from
the Red Arrows' original home of RAF Fairford, just a few miles
away from the team's then base at RAF Kemble. In 1968 the Red
Arrows had first been taken on a tour around the British-French
Aérospatiale/BAC prototype of Concorde, a new breed of aircraft
inspiring such excitement that no wonder the legendary aviation
photographer Arthur Gibson was keen to get a shot of the Red
Arrows in formation with Concorde 'before the French beat us to
it'. Team leader Ian Dick was similarly eager, knowing that such an
event would make for a historic moment in aviation, not to men-
tion great publicity. He knew John Cochrane, the Deputy Chief
Test Pilot for Concorde, and called him to see whether they could
work something out. On discussing it with the Chief Test Pilot,
Brian Trubshaw, they agreed that the Red Arrows could fly with
Concorde, but only at the end of a test flight as they headed back to
Fairford – with the proviso that this would be a one-off occasion.

A day was chosen, a time was set and they discussed and
briefed on their rendezvous. The plan was that the Red Arrows
team would get a call from air-traffic control when Concorde was
on its way home, at which point the team would get airborne and
would loiter just north of Kemble as Concorde approached Fairford. As Cochrane – who would be captaining Concorde for that
part of the day – put it: 'If I happen to see you in front of me, then
it would be churlish of me not to fall in behind you.' Ian Dick
had informed his chain of command about the planned event the
night before, although admittedly he presented it as more of a fait
accompli rather than putting in a specific request, fearing it might
not happen otherwise.

Ian describes the day's events: 'It was late when we got the call
to get airborne – after 4 p.m. on a dismal day in April. The Gnat

did not have a lot of fuel and we had to orbit for longer than I anticipated. I was getting a bit anxious about our fuel state when John came up on the frequency, and we were able to finesse the join-up. At the appropriate moment I set the team up on an easterly heading at one thousand feet in "Big Nine" formation.' At the same time the Team Manager was in another Gnat with Arthur Gibson, the photographer, in the back seat, and while doing a barrel roll over the formation, Gibson managed to get some fantastic images of the one-off flight. The team learned afterwards that the Concorde team had returned from a supersonic test flight over the Bay of Biscay, flying at more than 1,000mph, and that John Cochrane and Brian Trubshaw had been wearing bulky flying suits and astronaut-like helmets – quite a feat when flying a supersonic aircraft in such close formation with nine other jets.

The iconic photography of the team flying alongside Concorde in 1973 would be seen around the world, ably demonstrating the prowess of British aviation. The team would actually go on to fly with Concorde several times over the years, including with the Reds' new Hawk jets in 1980, at the RIAT in 1985, in 1996 for the fiftieth anniversary of Heathrow airport and in 1999 to mark the official opening of the new Scottish Parliament. In 2002 the team flew with Concorde for the final time as part of the twenty-seven-aircraft Queen's Golden Jubilee Fly-past, flying down the Mall and over Buckingham Palace. As Concorde was retired from service only a year later, the event would prove especially poignant, for the superb feat of engineering and design it represented. The partnership between the two nations of Britain and France, however, would not be forgotten, with the Red Arrows and their French counterparts, the Patrouille de France, flying in the shape of the iconic airliner at the RIAT in 2019 to celebrate fifty years since Concorde's first flight.

Over the years the team has also flown with the Avro Lancaster of the Battle of Britain Memorial Flight and with the Avro Vulcan (see first plate section), as well as a huge range of aircraft including an English Electric Canberra, Supermarine Spitfires, Hawker

Hurricanes, SEPECAT Jaguars, Panavia Tornados, Eurofighter Typhoons, the F-35 Lightning II, F-117 Nighthawk, P-51C Mustang, British Airways airliners 747s, 757s, 777s and the Airbus A380, the Airbus Military A400M, the Lockheed C-130 Hercules, and a Virgin Atlantic and Virgin Galactic 747 over Mojave in the United States, the last being a jet used to launch satellite technology.

A particularly standout mixed-formation flight for me occurred in 2015, when we led an Avro Vulcan V-bomber for the final time at the RIAT and then at the Southport Air Show. The Cold War-era aircraft, which once flew out of RAF Scampton, was in its final year of operation and the event marked the passing of an icon, one that had played a significant role in the history of British aircraft and was the undisputed highlight of air shows around the country. I remember the impact that the Vulcan V-bomber's power and size had on me as a child when I saw it take off and perform at RAF Waddington. Decades later there I was, the leader of the Red Arrows, watching the last airworthy Vulcan of the famous V-bomber fleet, XH558, named *Spirit of Great Britain*, repositioning so that it could join us. With ex-Red 11, Bill Ramsey, flying as co-pilot in the Vulcan, the plane's beautiful delta-shaped wings nestled brilliantly into our V-formation – a sight that Red 10, Mike Ling, described as being 'forever etched in my memory'. In leading the formation, I was so focused on delivering the fly-past safely that, on landing at RAF Fairford and taxiing out, I was quite overwhelmed to see people in tears. It was a reminder of how emotive aviation is in Britain, and how the Red Arrows can really help to pay tribute to such historically important aircraft – and it's a great honour for the team.

With its ability to fly in formation with jets as fast as Concorde, as well as the slower piston-engine aircraft such as the Lancaster, the Hawk really demonstrates its immense versatility. However, safely achieving a mixed-formation fly-past takes a lot of coordination – not to mention considerable planning and preparation, usually months prior to each event. Our job at the Red Arrows is to build a safe plan and to assure our chain of command that the given task includes well-thought-out contingency plans. If we are flying in formation with another aircraft and crew who aren't used to this type of aerial activity, we very much take the lead, and our planning process encompasses: the manner in which we meet (join) each other in the air; the speed at which we will conduct the fly-past and how to control the speed as a mixed formation; which radio calls will be made; what heights are safe to join and fly the route together; and what to do if an emergency situation develops for any member of the formation. The plan on paper is one thing, but making sure that our counterparts know how to fly it is another, so we also conduct a face-to-face briefing to ensure that everyone understands their role.

Prior to flying with the Red Arrows in formation, commercial airline pilots will fly the mission over and over again in a simulator. The time and resources spent on each fly-past ensure not only that it is flown correctly and safely, but also that each event has sufficient impact. Key to this is capturing the correct imagery to immortalise the moment. This might be achieved by a photographer in a helicopter hovering at precisely the right location, height and time – as we do for the Queen's Birthday Fly-pasts – or Red 10 and our own photographer will 'chase' the formation, lining themselves up with the jets and the scene, so that they capture the perfect image.

A ROYAL FAVOURITE

Over its fifty-eight-year history, the team has performed countless fly-pasts marking key national moments, from the Royal Tournament (once the world's largest military tattoo) in 1979, and the end of the Falklands conflict in 1982, to the fiftieth anniversary of 617 Squadron's famous Dambuster Raid in 1993, the Queen Mother's 100th birthday in 2000, the Great North Run and Commonwealth Games in 2002, and the FA Cup Final in the new Wembley Stadium in 2007. Mixed formations with Lancasters or Spitfires of the Battle of Britain Memorial Flight alongside the full range of modern military aircraft highlight both the important role that the RAF has played in the history of Britain and its ongoing role in defending the skies with Eurofighter Typhoon forces on Quick Reaction Alert, while projecting Britain's power and influence around the world.

The Red Arrows are often asked to appear at national events where a member of the Royal Family or other dignitaries are present – the appearance of the red jets in formation adding a final flourish to the occasion and, of course, making for a good photograph. Such was the case in 1981 when Her Majesty the Queen officially opened the Humber Bridge connecting the East Riding of Yorkshire with North Lincolnshire. It was, at that time, the longest single-span suspension bridge in the world and a real engineering achievement. The opening ceremony, held on 17 July, included a prayer of dedication by the Archbishop of York and a fly-past by the Red Arrows. According to the *Guardian*, the Archbishop of York had added an extra verse to his prayer of dedication, with the result that the Red Arrows performed their fly-past right in the middle of it.

The length of the Archbishop's prayer, however, probably had little to do with the early arrival of the Red Arrows, as Brian Hoskins, who led the fly-past that day, remembers: 'Some six months earlier, an army officer in charge of the whole event visited Kemble to brief me on our part. The basic plan was for us to fly south-to-north along the bridge as the Queen cut the tape. During the months leading up to the event I was sent changes to the plan and timing. These, especially the timing, concerned me, so I decided to have our own plan: we would arrive early and enter a Holding Pattern to the south. One of our team would be on the ground by the bridge with a radio, watching proceedings, so that he could confirm that things were running to schedule. On the day we entered the Holding Pattern in good time and I called our lookout on the bridge several times, but he did not respond. I knew that the photographer Arthur Gibson was also near the bridge in a helicopter, so I called him next and asked, "Arthur, have you got a good view of the bridge and, if so, does it look like they're running on time?" He answered, "Yes, it looks like they're on time." The only option we had was to fly over the bridge, trailing the red, white and blue smoke on the last time I had been given, which we did precisely. As we passed the northern end of the bridge, our lookout came on the radio: "Boss, you've come too early! Wait over Hull and I'll call you and we might be able to rescue this." We did as he said and he called us back as the Queen approached the tape. We flew up the river, and looped over the bridge into a downward bomb burst.'

While the event hadn't gone entirely to plan, the crowd at least got to see the Red Arrows perform, not once but twice over the great Humber Bridge, much to their delight. Brian Hoskins, still a little unhappy about the occurrence – the Red Arrows naturally pride themselves on their precise punctuality – was later informed that the vehicle in which the Queen was travelling had broken down due to a problem with its water pump, which, unbeknownst to the Red Arrows above, had added a seven-minute delay to the arrival of the royal entourage. The error lay with the faulty radio communication (and the water pump) and not, it would seem, with the Archbishop of York.

Not all fly-pasts are announced, however, as was the case on 6 July 2005 when the Red Arrows planned to do a surprise fly-over of London after the announcement of which global city would host the Olympic Games in 2012. Jim Turner was Red 5 that day and remembers the event: 'We launched out of Scampton, with Dicky Patounas as leader, and I was one of the wingmen. The idea was that we'd fly over London if we got the call on the radio with the relevant code name, but if we were given a different code word, meaning it hadn't gone well for the London bid, we'd turn around quietly and go back to Scampton.' It all worked out in London's favour of course, and 2012 would see the team perform a large number of high-profile fly-pasts – and not just for the Olympics.

On 19 May that year the team joined a fly-past featuring every aircraft from the RAF over Windsor Castle, to mark the Queen's

Diamond Jubilee, featuring a variety of big multi-engine and fighter aircraft, helicopters, Eurofighter Typhoons, the Battle of Britain Memorial Flight, Hercules, a VC10 leading a pair of Tornado GR4s, along with the more unusual formation of twenty Tucanos forming the number '60' in the sky, and twenty-seven Hawks forming 'E II R'. The Red Arrows in a special Deep Diamond Nine formation followed behind at the back, with the team working hard to maintain its position in the formation, despite considerable vortices coming off the wings of the multiple aircraft in front of them. The Reds made up the fly-past's grand finale and received the biggest reaction from the crowd, after what had been a challenging season for us, following two accidents among the team.

Later that same summer we carried out multiple fly-pasts across the UK to mark the opening of the 2012 Summer Olympics on Friday 27 July. We flew over Edinburgh at exactly 12.33, over Belfast at 13.01, over Cardiff at 16.21 and finally, at 20.12 – to represent the year 2012 – we flew over the Olympic Stadium in London, prior to the start of the Opening Ceremony. There were also further fly-pasts over London on 10 September, to mark the end of what had been a historic and golden summer for Great Britain in the Olympic and Paralympic Games.

That year marked future leader Martin Pert's first season with the team, flying in the Red 2 position. He remembers the build-up to 27 July and the intense sense of excitement for the whole team. They were fortunate on the day to have great weather as they approached London, and Martin describes it being 'an almost eerie evening, with a gorgeous sunset and a very high cloud, which set this golden light across London that evening. There was also very little air traffic over London, as a lot of it had been cleared or restricted, and the radios were really quiet and often silent.' Red 1, Jim Turner, also remembers the fly-past being a relatively easy one to lead: 'Unusually for the city, there were no "bumps" as it was such a beautiful still evening, which is one of the things you pray for when you're flying over London.'

The team reached the Olympic Stadium and crossed over it in a second, accompanied by the flashes of thousands of cameras, with the coloured smoke lingering in the still of the night. 'Seconds later,' recalls Martin Pert, 'we were over Hyde Park, where Duran Duran were playing and, as we went into a left-hand bend, I could see a large TV screen and we were on that screen, trailing the red, white and blue. It brings up the hairs on the back of my neck even thinking about it.'

THE QUEEN'S BIRTHDAY FLY-PAST

Usually taking place on the second weekend of June, the fly-past for the Queen's birthday is an annual event for the Red Arrows. On the ground, the big day begins with a military parade from Buckingham Palace in central London down the Mall to Horse Guards Parade, where the Queen attends the Trooping of the Colour. The Royal Family then gathers on the balcony of Buckingham Palace to watch an RAF fly-past. Flying 1,000–1,200 feet above the ground, the fly-past consists of a fleet of RAF aircraft, past and present, with the Red Arrows bringing up the rear to close out the show.

The year of 2015 was my first season as team leader and the fly-past that year was set to include a fleet of thirty-one aircraft, including Chinooks, a Puma and Sea King helicopters, a Dakota, Spitfires, Hurricanes, Tornados, Typhoons and a Hercules. Televised, and with the might of the combined armed forces looking on – not to mention thousands of spectators, plus the Royal Family – it's a high-profile fly-past for the Red Arrows. It's also a pressurised event for any new Red 1, as former leader Martin Pert remembers: 'You don't want to be the one flying over Buckingham Palace late or at a jaunty angle – you really feel the responsibility.' When I was leading the team as Red 1, I vividly remember feeling all the same pressures, and the 2015 fly-past would prove to be one of the most challenging I had ever done.

From a team leader's perspective, monitoring weather forecasts prior to a big event becomes something of an obsession. I might be engaged in a work meeting or on a family day out, but a quick peek on my phone at the latest weather forecast would see my mind wandering into contingency-planning mode, which made mental separation from the work environment quite challenging. The lead-up to my first Queen's Birthday Fly-past was no different, and three days before, it was becoming obvious that the weather was going to be particularly poor in the London area, with low cloud and rain forecast. The original plan was that we would take off from RAF Scampton on the day of the fly-past and refuel at Biggin Hill in south-east London, before launching into the fly-past for the Queen. However, with forty-eight hours to go – fearing that the weather would be so bad in the morning of the fly-past that it could jeopardise our chances of even landing at Biggin Hill, let alone making the fly-past – I decided that the team should fly down to Biggin Hill a day early, so that at least we'd be *in situ* and ready to launch if the weather took a turn for the better. Being flexible and changing plans at the last minute are a constant of military life, but this call came with some consequences, as it meant removing a day of leave for the team, which I didn't feel great about. But when it came to delivering a high-profile event, it would turn out to be the correct decision.

As predicted, the weather was poor on the day of the fly-past and the clouds were so low that many airports in the South-East were completely fogged out. As we arrived at Biggin Hill from our accommodation just a few hours before take-off, I could tell there was a sense of unease amongst the team. The usual banter was low-key and the pilots seemed more tense than usual. The problem we faced was that the weather forecasters had predicted the weather would improve at some point that day – but we didn't know when. This made decision-making very difficult for all the flying elements involved. The decision came in that we were to continue, and if a cancellation were to happen, the call would be made at the very last minute. So we conducted our usual pre-flight routine. During the brief I had to elaborate on what all the contingency plans might be, if the weather was still poor, as we still didn't have a legal-diversion airfield declared. In such a scenario the team leader must appear calm and in control, even if on the inside there are nerves and angst – now is the time to 'put your face on' and demonstrate to the team that you will lead them with total confidence, no matter what you are truly feeling.

At the very last minute before putting on our flying kit, an airfield in Norfolk was declared our diversion destination, should we need it. This meant that we could at least progress to take-off, but would we make it to the Palace? I wasn't so sure. Lining up on the runway at Biggin Hill, ready to take off, we still didn't know whether we would get to London or not. I had briefed all our contingency plans and kept visualising them on a loop as we slowly taxied to the runway. By now my nerves had diminished. I was fully in the moment and, even though we might disappoint Her Majesty and the assembled crowds, I was mentally telling myself to make safe decisions, no matter how important our audience might be.

As we were still lined up at Biggin Hill, with forty-five minutes to go before the fly-past, air-traffic control informed us that all the other aircraft in the fly-past fleet had cancelled – the decision being that it was safer for us to have more flexibility in finding a route into London without having to get together with a long train of other aircraft. Sitting there on the rainy runway and having willed myself to dissociate from the high profile of our royal audience, I remember thinking: 'Okay, so if we don't make it there, then not a single plane will show up for the Royal Family waiting on that balcony.'

I took a very deep breath and decided that it was now safe to get airborne, reminding myself not to succumb to task pressure over sensible decision-making. Once we were all airborne, it proved phenomenally challenging to find a gap in the clouds and rain in which we could see each other properly. I eventually found a route south of Clacton, near Colchester in Essex, where there was a pocket of clearer air, and because we cut short the rest of the route we arrived early and had to circle around our position – nine aircraft split into two sections, simply going round and round. There was a golf club below and I remember seeing golfers looking up at us, no doubt wondering what on earth we were doing as they tried to keep focused on their game, prompting me to radio through to Red 6, Mark Lawson, 'I think we're going to disturb their handicap today.'

That brief moment of levity at least took the edge off the pressure that we were all feeling, as it was still uncertain whether we would make the route into London. Having made the decision to push on, I judged the correct timing to start the approach to ensure that we were there to the second, and positioned our formation on the correct track. The cloud base was high enough for us to progress safely enough, but the real issue became reducing visibility. With more rain beating down on the canopy, I mentally prepared to action the abort, at which point we would immediately have pulled up and away from the ground – which we could no longer see – and tell air-traffic control to provide us with a radar service.

We were now about ten miles away from London, a far better place to abort than over the air space of the city, and I remember that my right hand was about to put pressure on the control stick to pull up when we suddenly burst out of the cloud and the visibility went from about three to four miles. So we stuck with it, keeping the formation at the same altitude. There's often a bit of a sixth sense between Red 1 and the leader of the back section, Red 6, and Mark told me later that he could sense I was about to go up and was similarly primed to go with me. Until this point Mark had done an amazing job of keeping the rear section in good order, and safely spaced behind me and Reds 2 to 5. We talk a lot about the notion and importance of trust amongst the team, and this occasion was definitely a good example of the bond between Reds 1 and 6: Mark trusted that I would make the correct decision at the correct time and he would go with me, whatever the choice.

Once we were about five and a half miles from London I could see Stratford, the location of the Olympic Stadium and a convenient feature on the landscape, marking the initial point for the Queen's Birthday Fly-past, leading straight down towards the Mall. London itself was murky, but the weather was steadily improving and I knew then that we were going to make it. Moments later – and with an enormous sense of relief on my part – we were flying over central London. With the smoke on, we were bang on time, and we flew down the Mall in our usual V-formation and over Buckingham Palace, albeit with no other aircraft ahead of us. Throughout the whole flight my Circus 1 engineer, Flt Lt Marcus Ramsden, was in the back seat, keeping very quiet, as he could hear how difficult things were for me up front, with occasional huffs and gruffs audible amongst my usual radio calls. On landing back at Biggin Hill, I don't think I've ever seen Marcus look quite so elated, but he was of course carrying a great deal of the stress of that day as well. For me personally, it was a huge milestone: not only was it my first Queen's Birthday Fly-past as Red 1, but to do it in those conditions and be the only aircraft to make it through felt like a remarkable achievement.

After an hour of flying, I felt like falling out of the aircraft at Biggin Hill from utter relief and exhaustion, which as leader you tend to play down a bit, as you want to be there for your team. Fairly soon after landing I received a phone call from Air Vice-Marshal Gary Waterfall, the most senior officer responsible for the organisation of the Queen's Birthday Fly-past. Hearing his voice, I thought to myself that I was either going to be congratulated on a job well done or get the sack for deciding to press on in less-than-ideal conditions. Having heard that we were airborne, he said that he had given the call for the Royal Family to stand on the balcony, while he and other military dignitaries stood at the top of New Zealand House, where they had a bird's-eye view of the fly-past. It turned out that he was just as jubilant as we were to have made it: 'Monty, I cannot tell you how relieved I was to see your nose lights coming out of the murk. Thank you very much.'

While the Queen has seen the Red Arrows do a fly-past down the Mall on an annual basis over the last fifty-six years (bar the odd cancellation), she had never seen them do a display until July 2013. A few months earlier the team had been invited by the Custodian of Windsor Castle (the former Air Marshal Ian Macfadyen) to have a tour of the castle, during which, quite unexpectedly, it was invited to meet the Queen. Jim Turner, who was team leader that year, remembers: 'We filed into this beautiful hall, at the end of which we could see the Queen's Corgis yapping. A butler appeared (who we recognised from the video at the London Olympics featuring the Queen and Daniel Craig, as James Bond) and invited us into her private living room – and it was talking to us then that she mentioned she had never seen a display.' Eager to set this right, the team flew a practice display for her at RAF Marham, close to the Queen's country retreat of Sandringham in Norfolk. During the display Mike Ling was on hand to talk the Queen through the show, and afterwards the team was thrilled to be invited to an informal lunch with Her Majesty.

The Red Arrows were also invited to Windsor Castle a few years later, in March 2019. While chatting with team leader Martin Pert, the Queen mentioned that the last time the Red Arrows had performed over Windsor Castle, her great-granddaughter had taken a photograph. 'She opened her handbag and pulled out some six-by-four prints – clearly very proud of them – one of which was a beautiful shot of the Red Arrows taken during the Diamond Jubilee celebrations of 2012,' remembers Martin. 'I was Red 2 in 2012, so I knew that I was flying in that formation, so obviously I couldn't wait to tell my mum that the Queen kept a picture of me in her handbag.'

CENTENARY CELEBRATIONS

As the oldest independent air force in the world, the Royal Air Force celebrated its centenary on 10 July 2018, and the Red Arrows team, with Martin Pert as leader, would spearhead those celebrations and provide the finale for a fly-past over central London. It featured 100 aircraft – one for every year of the RAF – including historic Battle of Britain Memorial Flight aircraft, as well as three new F-35B Lightnings, and twenty-two Typhoons flying in a '100' formation, with the Red Arrows drawing everything to a close. To honour special events and commemorations, the team often revises or designs new shapes and manoeuvres for displays, and this year proved no exception. Alongside such crowd-favourite manoeuvres as the Tornado and the Phoenix, the term performed the Centenary Split at the start of the second half of the display. This saw seven Hawks pull up at 4G and 420mph, climbing more than a mile high before splitting into a fan-like break, while the synchro pair conducted a 'flat split' beneath them. The centenary display finished with an impressively large '100' in the sky, created out of white smoke.

The Red Arrows make much of what they do look effortless, but in reality the '100' proved challenging for the back section of the formation who flew the shape, and Red 1 Martin Pert wasn't entirely convinced the team would be able to include it in the performance. As Martin recalled, '100 sounds simple as you would think that one jet simply needs to fly straight up, while another two jets fly loops. But when an aircraft loops, it doesn't fly in a perfectly circle shape, but more of an oblong. As the "100" moment was at the end of a display, the team had already used quite a bit of fuel and smoke, so that also dictated who would fly it, which ended up being Reds 6, 8 and 9. After considerable training, repeated practice and further finessing, assisted by feedback from Red 10 watching on the ground, they cracked it and we were able to use it as a fitting finale to the year's displays.'

Her Majesty the Queen has probably seen the Red Arrows more than most people in the country, and yet her face still seems to light up every time the nine red jets fly overhead. She has remarked that when standing on the balcony of Buckingham Palace, she feels fortunate to get the best view of the Red Arrows whenever they fly down the Mall: glimpsing first the white smoke streaking through the sky, then the red and blue colours, accompanied by a ripple of cheers from the crowd as the Reds cross over.

THE PRINCES AND THE PILOT

As Head of the Armed Forces, the Queen obviously has a great regard for and strong links with the Royal Air Force, and various members of the Royal Family not only hold honorary ranks with the RAF, but their own pilot's licences. The late Duke of Edinburgh held the rank of Marshal of the Royal Air Force and received his RAF wings in 1953. The Prince of Wales holds the rank of Air Chief Marshal in the Royal Air Force and, during his second year at Cambridge University in 1968–9, received flying training from the RAF, after which he trained as a jet pilot at RAF Cranwell and, during his naval career, qualified as a helicopter pilot in 1974. In 1969 Dickie Duckett, who was then Red 4, remembers Prince Charles in his flying suit dropping in to RAF Kemble, where the Red Arrows were then based. He was undergoing navigation training with the Commandant of the Central Flying School in a twin-engine piston aircraft. 'It was all very informal, we had a chat and coffee and off they went again,' recalled Dickie, showing that sometimes royal visits can be very low-key affairs.

A regular visitor to the team, particularly in the early years of the Red Arrows, was HM Queen Elizabeth, the Queen Mother. After 1960 she held the rank of Commandant-in-Chief of the Central Flying School, the parent unit of the Red Arrows. Brian Hoskins, team leader between 1979 and 1981, remembers meeting the Queen Mother after a display at RAF Leeming, and she appeared very knowledgeable about the team. In 1991 the Red Arrows did a special fly-past over the Queen Mother's Scottish residence at Birkhall on the Balmoral Estate. The Reds flew in formation with the Russian Knights aerobatic team, having visited Russia for the first time the previous year. Nine years later, to mark the Queen's Mother's 100th birthday, the team created a special fly-past of ten jets – one Hawk for each decade – over Horse Guards Parade in central London.

The Red Arrows had another royal visitor in 2008, this time Prince William, Duke of Cambridge. He of course has a background in the military, having served in the army from 2006, where he learned to fly Tucanos and Squirrel helicopters, receiving his wings at RAF Cranwell in 2008, before transferring his commission to the Royal Air Force to become a helicopter pilot with the RAF's Search and Rescue Force. His visit in 2008 was an informal one, away from the cameras, during which he got a chance to fly in the back seat of one of the Hawks and experience some manoeuvres of a typical display, which gave him a really good insight into what we do at the Red Arrows. While he no longer flies helicopters, Prince William is now Patron of the Royal Air Force Battle of Britain Memorial Flight and Honorary Air Commandant of RAF Coningsby, and in doing so he draws attention to the crucial role played by UK air defence while commemorating its important history. More recently, in December 2020, Catherine, Duchess of Cambridge took over from the Duke of Edinburgh to become Honorary Air Commandant of the Air Cadets.

In July 2016, and more in the public eye this time, Prince William and Catherine, Duchess of Cambridge, plus their son Prince George, attended the RIAT at RAF Fairford, where they met team members of the Red Arrows. As Red 1, I had met Prince William at a charity event the previous year, but this was the first time I experienced the huge entourage and cameras that invariably accompany the Royal Family at public events. The previous year, at the aforementioned Queen's Birthday Fly-past, the world's press had captured images of toddler George in the arms of his mother, waving and smiling as we flew overhead, so it was perhaps only a matter of time before he got the chance to see the Red Arrows up close.

We had long been scheduled to display at the RIAT, but we were given just twenty-four hours' notice that we were going to meet the Royal Family there, at which point our team organised everything to ensure that we were going to be at the right place at the right time. We met them outside, alongside our parked Hawks, and I initially talked to the Duchess and Prince George, who, at nearly three, was roughly the same age as my own son. As I knelt down to talk to him, the Prince was understandably quite shy. I asked whether he remembered seeing us before, and the colours coming out of the back of the jets. After the Duchess reminded him that he had seen us at his grandma's house (aka Buckingham Palace), we talked about the red, white and blue smoke. The Duke then came over and we had an amiable chat about parenting, almost as if we were two dads at the school gates, rather than a prince and a pilot in front of the world's press. There was, naturally, a bit of banter about whether Prince George was going to be a fan of fast jets or helicopters when he was older, with his father joking that it would definitely be the latter.

The most stressful time of the whole encounter – and in fact of the entire four days of ground engagements that week – came when Prince George said he wanted to get into the cockpit. I was more than relieved when the Duke, looking at me, asked if he should lift Prince George into it. Entirely safe as the whole thing was, it was still slightly unnerving to know that the third in line to the throne was sitting upon multiple rockets and explosives embedded within the ejection seat. The Duke was, of course, familiar with the layout of the cockpit and he and I talked the Prince through it – a real thrill for anyone who has never been in a fast-jet cockpit, whatever their age. After George got a little excited and put his hand on one of the instrument panels, William lifted him out really carefully, and I think my heartbeat then went from the 160bpm it had suddenly shot up to, back down to around 60bpm.

The whole time there was a huge bank of paparazzi taking photos and filming, and I was pretty amazed at how quickly the images shot around the world. The engagement was captured in *Hello!* magazine, was circulated in some fifty to sixty countries, and it wasn't long before I had family from Colombia, some of whom I hadn't spoken to for years, getting in contact to tell me there was a picture of me with the Duke of Cambridge and Prince George. While the Red Arrows have quite a bit of reach when it comes to generating publicity, it comes nowhere near the level the Royal Family can achieve, in terms of global footprint and influence. Nonetheless, at the time the UK was going through quite a difficult period: the referendum on membership of the European Union had happened the previous month, the Prime Minister had just resigned and the nation was in a state of flux. With all that going on, I hope that the Red Arrows – for aviation enthusiasts at the very least – provided a certain sense of continuity and reassurance, and a welcome distraction from everything that was happening in the news at the time.

The feel-good factor and the reassurance that the Red Arrows can sometimes bring to the nation were certainly felt during the 2020 season, when the world was in the grip of the Covid-19 pandemic. While many of the usual air shows and fly-pasts were cancelled, the team led by Martin Pert performed some high-profile fly-pasts, including for VE and VJ Day, marking seventy-five years since the end of the Second World War in Europe and the Far East. For VE Day on 8 May the jets trailed their signature colours over central London as well as over the Runnymede Air Forces Memorial. For VJ Day on 15 August the Reds flew over Belfast, but had to cancel fly-pasts over Edinburgh, London and Cardiff during poor weather, returning to Edinburgh nine days later to mark what would have been the Edinburgh Tattoo. While we minimised the publicity leading up to fly-pasts to avoid large crowds forming, we received a really positive response from the nation, and I don't think I've ever seen so many people waving up at us from their gardens.

FLYING WITH THE STARS

The first glimpse some people get of the Red Arrows may not be at an air show or fly-past but on their small screens, and over the years the team has featured in a variety of television programmes as well as radio series, and have been visited by an array of celebrities and elite sportsmen and women. One TV presenter, John

Noakes, got the full Red Arrows experience back in 1975 when he joined the team at RAF Kemble for his BBC *Go With Noakes* series. Having been an aircraft fitter previously, Noakes was familiar with the insides of jet aircraft, but was not so experienced at flying in the back seat, as he did with Squadron Leader Dickie Duckett. While flying a few loops and rolls in the Gnat – described by Dickie as 'like a racing car with wings' – Noakes was surprised to be given the controls, and even more surprised when he performed a 360-degree Twinkle Roll, after Dickie had told him to push hard left over on the stick. Noakes then experienced the rigours of a full sixteen minutes of display, as the team flew a variety of manoeuvres, ranging from the Concorde and the Feathered Arrow to a Vixen Break. He then joined the Red Arrows' ground crew, before accompanying the team manager in his helicopter while the team displayed at various events.

The following year Dickie Duckett in his Gnat went on to race the famous motorcyclist Phil Read on the runway of RAF Kemble; and over the years a number of famous motorcyclists have visited the team, from Barry Sheene in 1979 to Carl Fogarty and Troy Corser, as well as the TV racing presenter Suzi Perry in 2001. The Red Arrows team has long had links with the world of fast machines and racing events, regularly appearing at the Isle of Man TT, the British Grand Prix and the Goodwood Festival of Speed, along with various sailing regattas, including Cowes Week. In 1967 Formula 1 racing legend Graham Hill flew with the team, as did Nigel Mansell in 1989, after which he was surprised by the TV presenter Michael Aspel, who presented him with the *This Is Your Life* big red book on the airfield at RAF Scampton.

In 2013 Formula 1 racing drivers David Coulthard and Lewis Hamilton created quite a stir when they joined the Reds for the day at RAF Scampton. After medical checks and a briefing, both got to fly in the back seat of a Hawk and experience the first half of a display, flying in close formation with a range of manoeuvres. Hamilton described the G-force as being very different from that he experienced in a car. Shortly after his flight though, Hamilton was ready to jump into his F1 car and did a demonstration on the runway with one of the Hawks – with Sqn Ldr Mike Ling and David Coulthard aboard – flying above RAF Scampton, trying to sync their speed as they travelled. The amazing footage was seen by a global audience. Although it definitely wasn't a race, it would never really be a fair one if it were, as our Hawks can reach speeds in excess of 600mph – that's around three times the speed of a modern F1 car. Hamilton described the day as 'incredible' and 'the coolest thing I've ever done' – quite a compliment from a world champion F1 driver.

Over the years a huge variety of elite sportsmen and women from outside the racing world have also visited the team, including England cricketer Ian Botham in 1982; Olympic and world champion ice-skaters Jayne Torvill and Christopher Dean in 1984; England rugby player and pilot Rory Underwood in 1990; and Olympic swimmer Sharron Davies and Olympic sprinter Jason Gardener in 2002. In 2016 Olympic sprinter Adam Gemili visited RAF Scampton to participate in a 100-metre race with a Red Arrows jet on the runway for ITV's *It's Not Rocket Science*. At 100 metres on the ground, a Hawk has probably reached 30–40mph before accelerating to 140mph at take-off, with the result that Gemili was beaten. Afterwards he remarked: 'Racing against a British icon like the Red Arrows, I feel very lucky to be a part of that, and as a sprinter for Team GB it sets me up nicely to go and race in the Rio Olympics.' The parallels between elite sportspeople representing their country – the hard work and discipline that go into performing at the very highest level – and the Red Arrows are clear to see.

The BBC children's programme *Blue Peter* has also had a long association with the Red Arrows: John Noakes was a much-loved *Blue Peter* presenter when he flew with the team in 1975, and ever since that time a string of presenters have flown in the back seat of a Hawk, including Katy Hill in 1998 and Helen Skelton as part of her Red Nose Day 'Seven Challenge' fundraising effort in 2013. Skelton flew with Red 10 Mike Ling, who was immensely pleased to receive a *Blue Peter* badge for his efforts. More recently, in 2018,

Blue Peter presenter Lindsey Russell visited the Red Arrows team as they were starting Springhawk training in Greece. Red 10, Adam Collins, flew with her in the back seat of his jet alongside the other nine aircraft, so that she could see the formation perform a few manoeuvres. Then they set off for a few solo aerobatics, with Lindsey taking control of the aeroplane, just as John Noakes had more than forty years before. To mark the sixtieth anniversary of *Blue Peter* and the RAF's centenary, the show also ran a competition for viewers to help design new Red Arrows helmets to be worn at the RAF's centenary fly-past that same year. The show had a huge response to the competition, and I had the honour of helping to select the winning designs for the three helmets, which were eventually worn by Red 1 Martin Pert, Red 8 Matt Masters and Red 9 Mike Bowden during a fly-past over central London.

Various well-known names from the world of show business have also visited the Red Arrows, including Scottish-born movie actor and producer Gerard Butler, who flew with the Reds during their tour of North America to help highlight the creative industry's links between the UK and US. Back in 1991, actor Sir David Jason also visited the team – a qualified helicopter pilot and Honorary Vice Patron of the RIAT, he was a long-time supporter of the Red Arrows. And in June 2011 Queen guitarist Brain May dropped into Scampton along with West End star Kerry Ellis, as part of fundraising efforts for the RAF Benevolent Fund.

During my time as team pilot, BBC's *MasterChef* came to film one of their programmes at RAF Scampton in late 2014. The contestants were set the task of cooking two menus, fittingly titled 'red' and 'blue', for RAFAT while the team went about a normal day of training. As the meals were cooked and served in one of our hangars, the contestants had to work about eighty feet away from the back of our jets starting up for training – making for trying conditions and a whole lot of noise. The presenters Gregg Wallace and John Torode, who were part of the two-day filming project, really enjoyed their time with the team, and we discussed the similarities of a busy kitchen with life at the Red Arrows: close teams working to the clock; lots of pressure to perform in difficult conditions; not to mention a stressed-out head chef/Red 1 up front. Some seven million people watched that *MasterChef* programme, which proved to be the most successful media moment for the RAF (and probably UK Defence) that year – and the salmon that I had for lunch certainly powered me onto the third sortie of the day.

In the summer of 2019 the Red Arrows were particularly honoured to host two very special guests: British and Canadian astronauts Tim Peake and Chris Hadfield. Peake flew in a practice show at RAF Fairford as the team prepared for the RIAT in July 2019. During his twenty-minute flight he experienced at first hand a Red Arrows display, including the Apollo manoeuvre, a shape that marks the fiftieth anniversary of the Moon landing. Tim, the first British European Space Agency astronaut and a trained helicopter pilot, described the experience as a 'fantastic flight ... I fulfilled a boyhood dream today, I really have.' High praise indeed from someone who has travelled in space. Just over a month later, during the team's North American tour, Squadron Leader Martin Pert took the Canadian astronaut and former fighter pilot Chris Hadfield in the back of his Hawk over Toronto in a unique mixed formation with the Canadian Forces Snowbirds. The flight formed part of the Red Arrows' ongoing work to inspire young people to consider careers in STEM subjects.

During the time that RAFAT was flying astronauts in the Hawks' back seats, I was living in Australia and was lucky enough to be the RAF representative studying for an MA at the Australian War College in Canberra, having finished my tour with the Red Arrows at the end of the display season in 2017. Six months earlier, on a day when my family and I were packing up and clearing out our service quarters before heading to Australia, I received an unexpected phone call from the station commander at RAF Coningsby, where I had been working. 'Congratulations,' he announced. 'You've been awarded an OBE.' On putting the phone down, I didn't know what to do with the information and carried on cleaning behind a radiator, before eventually sitting down and trying to take in what had come as a bolt out of the blue. I was told it was in recognition of my three years as team leader, in particular getting the team to China and then straight into another year of further engagements in the Middle East. I was, of course, shocked. There have been others at the Red Arrows who have received the honour – John Rands before me, one of my predecessor Red 11s; Group Captain Martin Higgins received a CBE; former team leader Jim Turner and my closest colleague, Mike Ling, had both received an MBE – but it was not something I'd ever dreamed would come my way. More recently I was immensely pleased to hear that Kirsty Murphy and Martin Pert had been similarly honoured, the two former Reds being respectively awarded an MBE and OBE in recognition of their service.

Luckily I was able to defer attending the investiture ceremony at Buckingham Palace until February 2020, once I had returned to England from Australia. This time my view of the Palace was not from the cockpit of a Hawk T1 – three seconds and it's gone – but at a more leisurely pace, as my family and I walked through those grand entrance gates. HRH Prince Charles was awarding the honours that day – someone who's seen the Red Arrows a fair bit during his life, from the deserts of Oman to a rain-sodden RAF base in Gloucestershire, and now in the rather more lavish surroundings of a state room at Buckingham Palace.

A NATIONAL HERITAGE

The occasions that we mark at the Red Arrows, the great national events and commemorations, plus the incredible cities and landscapes that we fly over across the globe, from Dubai and Toronto to Karachi and Athens, make for a string of memories that are treasured by every pilot and team member who has flown with the Reds. The Opening Ceremony of the Olympics, a fly-past over New York Harbour and flying a formation with Concorde all make for a striking photo and serve to fly the flag for the UK and the Royal Air Force, and generate publicity across the world. Those headline-grabbing events are what we strive for at the Red Arrows, but there are also countless other moments, not so much in the public glare, that are just as significant or meaningful to the team at RAFAT.

For some people, it's getting the news that they're in the team – Kirsty Murphy learning that she'd got into the Red Arrows, in the very same room at RAF Marham where she first saw her father after he had been shot down and taken prisoner during the Gulf War. Or it's that late-afternoon display in cloudless skies flying over the Lake District when the team is working to perfection. For JEngO and Circus 1 Ben Ireland, it was getting the job that he'd aspired to since joining the RAF, and solving that niggling problem in a random airfield in the middle of nowhere. For many, like Ian Dick and Martin Pert, it's meeting the challenge of leading a formation of single-engine jets across the Atlantic – there's no particular fanfare when they arrive on the shores of North America, but the sense of achievement is as great as, if not greater than, when flying over Victoria Falls.

For many team members the standout moments are when they can perform not only to crowds of hundreds of thousands, but also to small groups of family and friends, or at air shows where they remember standing as young children, gazing up at the awesome antics of the jets above. Longest-serving Red Arrows pilot Mike Ling felt he had truly fulfilled his boyhood dream when, in July 2009, he did a fly-past with a Boeing 747 at Biggin Hill – an air show that he'd first visited as a three-year-old, returning every year to queue up to meet the Red Arrows. His grandfather had been a Lancaster pilot, and for Mike to have his family and friends there watching him made for a very special day. For Kirsty Murphy, it was getting through her first public display at RAF Brize Norton in 2011, when she suddenly realised that the training really worked. When she landed she was mobbed by four young girls: 'That's when I realised that being the first female Red Arrow was really something – it's bigger than me.'

Sometimes it's those days when everything comes together as a team: the nerves melt away and those months of training kick in, and it's almost as if the aircraft is flying itself, the formation entirely in sync, turning gracefully as one 'locked-together welded wing'. It's knowing your aircraft so well that, occasionally, you and the team come within reach of perfection, up there in the rarefied air of the skies. You never know when those moments will come, but it's what you yearn for as a pilot.

Afterword

The experience of being in the team – what it is to be a Red Arrow – is unique to every individual, being shaped not only by the context of the time and the culture within the service, but also by the extreme highs and lows that occur during one's time on the team. I am but one voice (out of the many hundreds of people who have devoted themselves to the Red Arrows) and I could never speak for all those individuals. My aim is to shed some light on my experiences, alongside the voices of those who have contributed to this book, in the hope that I can give a flavour of what life on the team is like.

The Red Arrows are quite unlike any other military organisation. The sights, smells and sounds are unparalleled, principally because team members engage with so many people around the world: with the general public, politicians, royals, and with people across the generations and the full spectrum of society. In terms of the flying, our pilots are trained to fly at 100 feet above the ground and in opposition passes in front of millions of live spectators, all of which is very distinct from front-line life. It is why many pilots in the RAF are keen to join the team, as it gives them a chance to broaden their aviation horizons and learn a very specialist kind of flying. In the same way, our engineers – especially those who are selected to be Circus members – spend hours in the rear seat of a fast jet, an experience that is vastly different from most of their peer group working on other squadrons.

And yet the challenges and relationships that form within the team are similar to those in any other military organisation. We depend on trust, built upon the professional and emotional capability to lead people properly. Members across the team must at times make robust and unpopular decisions. Empathy, compassion and tenacity are characteristics that are synonymous with any successful military unit writ large, and the Red Arrows are no different. We may wear red and blue flight suits, but underneath the special display coveralls we are ordinary service folk.

I've always held the view that we, the team, must never regard ourselves as elite, as the best in the world; that is for others to judge, not us. While we take immense pride in the reputation built by the many who have come before us, we continue to strive for that perfect performance – and, at the high bar we set, such performances are few and far between. Every ex-Red I've spoken to remembers that one truly perfect show, whether it was a beautiful summer's day in the Cotswolds, a scorching Huntington Beach in California or looping and rolling in the electric atmosphere of Zhuhai in China.

And, in that pursuit of excellence, we never forget those who have lost their lives on the team. Tragedy has led to the most painful ways of learning lessons, but we in the RAF (not just the Reds) do our very best to ensure that we don't have to re-learn lessons from the past, by constantly assessing and ensuring that we are working within as safe an operating environment as we possibly can.

Perhaps the final thing – the only thing – I can say on behalf of any pilot, engineer and all the support trades that keep the Red Arrows team in the skies is that we are phenomenally lucky ever to have had the opportunity to serve in this brilliant squadron. To be given the honour to represent our service colleagues in such a dynamic way, and to be ambassadors for the UK, is a truly life-changing experience – and one that I hope many future generations have the opportunity to experience for themselves.

Acknowledgements

Creating this book has been a collective effort by past and present Red Arrows team members who have been willing to share their personal contributions, particularly Air Cdre Dickie Duckett (retired) and Gp Capt Brian Hoskins (retired). I wish to thank them for lending their time and offering valued insights into the enduring evolution of the Red Arrows. My thanks to Emma Marriott, and to Jess Ballance and the Penguin Random House UK team, and also to Andrew Morton, for all their invaluable support along the way.

To our public audiences and dedicated fans, we treasure your unstinting support throughout the highs and lows – thank you for being the fuel that continues to propel this team, year after year.

I also wish to thank my parents, who gave me the best possible chance to reach my dream – and for funding a one-off flying lesson on my sixteenth birthday at Biggin Hill Airport. I will never forget the elation I felt at touching the controls of an aircraft in motion for the very first time.

Finally, to Angie and our boys, for daring to be with me for the smoothest and bumpiest of landings.

Glossary

'100' shape: a manoeuvre that creates a number 1 and two zeros in the sky using smoke

Ace: a military aviator credited with shooting down five or more enemy aircraft during aerial combat

Aileron roll: a full 360-degree revolution about the jet's longitudinal axis. When done correctly, there is no significant change in altitude from entering and exiting the manoeuvre and the aircraft exits on the same heading as it entered. Commonly confused with a barrel roll, the aileron roll's centre of rotation is much closer to the aircraft, resulting in a tighter roll spiral

airway structure: a mass network of specified high-altitude air routes that ensures everyone gets from A to B safely

ahead of the jet: an RAF phrase for the ability to organise and structure the mind to think ahead. It's not about making lots of decisions at speed – a good fast-jet pilot will prioritise their thoughts in order to simplify their environment so the brain can make really good decisions at pace. This is a universally important skill-set, whether you are flying at twice the speed of sound in a Typhoon jet over the North Sea, or leading nine Red Arrows across the UK to arrive at a display location exactly on heading, on height, at the right speed and at the right time, to the second

A-loc: an abbreviated term meaning 'Almost loss of consciousness', a condition when an individual suffers profound cognitive and functional impairment after experiencing G-force

Angels: RAF shorthand for a thousand feet in the air

Apollo: a triangular formation named in honour of the 1969 Moon landing

armourers: weapons technicians who maintain the aircraft's ejection seats, explosive cockpit glass canopies and fire-suppression systems

Arrival loop: a vertical manoeuvre that makes for an impressive start

Arrow: the closest formation reference flown between the Red Arrows aircraft with Reds 6 and 7 in the stem, forming an Arrow shape

barrel roll: a graceful 360 roll that uses the vertical, horizontal and lateral axes, shaped as if the aircraft were flying on the inside of a traditional barrel

Battle: the classic symmetrical V-shape of the Red Arrows

Big Vixen: one of the core triangular-shaped formations

Blackbird: a manoeuvre during which the whole team moves into one of the longest and sleekest formation shapes, in which Reds 1, 6 and 7 form the stem, leaving the rest to outline the elegance of the Blackbird SR-71 reconnaissance aircraft

The Blues: Red Arrows engineers, technicians and support personnel, named after the blue suits they wear

Bomb Burst: when the aircraft peel away from the formation in different directions at the same time

Boomerang: a formation that sees the synchro pair run in, in front of the crowd, split, pull up into a loop, roll over and cross over

buffet: a gentle vibration in the airframe of the Red Arrows jet

'the bumps': gusty conditions that cause the aircraft to move around by two to three feet, meaning that pilots have to work extra-hard to keep themselves in position

CAA: Civil Aviation Authority

Carousel: a manoeuvre in which the synchro pair performs opposition passes in front of the crowd, then turns and performs another further back, before finishing off in a circle at the front

Centenary Split: a revised version of the Palm Split, which sees seven Hawks pulling up at 4G and 420mph, climbing more than a mile high before splitting into a fan-like break

CFS: Central Flying School

check-in: the point at which every pilot in the formation confirms that he or she is primed and ready to go

Chevron: a flat V-shape formation

Circus: ten engineers and one photographer, chosen to fly in the passenger seats of the team's Hawks in transit flights during the display season to form an essential support team when operating away from base

Class A airspace: generally begins from 18,000 feet mean sea level up to and including 60,000 feet. Operations in Class A are generally conducted under Instrument Flight Rules and primarily used by higher performance aircraft, airline and cargo operators, et cetera

clear round: a chance to clear all fines incurred, if you perform a sortie without any mistakes or just one; if more than one mistake is made, you risk doubling the amount owed

Concorde: a formation depicting the iconic shape of the Concorde aircraft

contract: the period when each section of the Reds is on the display site and effectively owns the forty-five to seventy seconds of audience viewership; the changeover between Reds 1 and 6 is known as the 'contract handover'

control column: this is used by the pilot to steer the aircraft in pitch and roll. The pilot holds the control column in the right hand

creamie: RAF slang meaning the *crème de la crème* or the cream of the crop, playfully used to describe the highest-quality of pilot

Cutlass: a formation shape depicting a short, broad sabre sword

Deep Diamond: an adaptation of the Diamond Nine created for the Queen's Diamond Jubilee of 2012 and given a more three-dimensional feel, with aircraft stacked at different heights so that the Diamond shape can be seen from the ground as the team comes in and out, rather than only when the jets fly overhead

Diamond Nine: the iconic shape of all nine Red Arrows flying in formation to make a diamond shape

Display Directive: a guide on how to be a Red Arrow, chronicling the Standard Operating Procedures (SOPs)

display take-off: a take-off performed by all nine aircraft. The team is split into three sections all lined up on the runway with Reds 1-3 at the front. Reds 4 and 5 take off behind the lead section and quickly join them. Red 6 leads 7, 8 and 9 to join the lead section, forming a diamond shape

domestic factors: factors that centre on the immediate environment, for instance the weather or availability of jets

Enid: Reds 1 to 5, who make up the front and outer sections of the Red Arrows formation; so named after Enid Blyton, author of *The Famous Five* books

F5000: records that include every flying report from day one of a pilot's training

F700: a large folder in which a pilot 'signs the aircraft out', indicating that responsibility for it passes from the engineer to the pilot

Feathered Arrow: a formation shape representing an arrow with feathers at the tail (used in a bow and arrow). The 'feathers' are represented by Reds 8 and 9 (see Fred)

Finger Four: a Second World War formation consisting of two pairs of fighters, with each pair containing a leader and a wingman

flat display: this consists of fly-pasts and steep turns and will be flown if the cloud base is as low as 1,000 feet

Flt Lt: Flight Lieutenant

flying by ear: voice control of manoeuvres

formation flying: two or more aircraft travelling and manoeuvring in a synchronised manner

formation reference: the exact position an aircraft flies in relation to another aircraft – the reference is visually assessed by triangulating two points on the leader's aircraft so an exact position is maintained

Fred: another term for Feathered Arrow, but much easier to say during a radio transmission (see Feathered Arrow)

full display: this includes performing full loops without the danger of entering cloud at the top of a loop and will be flown if the skies are cloud-free, or at least clear to above 5,500 feet

gate speed: target starting speed

G-force: a measure of acceleration, felt due to the force of gravity. When performing an aggressive manoeuvre such as the Carousel, the synchro pair might experience 6–8G, which effectively causes the head and body to weigh six to eight times more than normal

Goose: a manoeuvre that sees Red 8 or 9 sprinting towards the five aircraft of Enid in Pyramid formation and crossing straight through the middle of them

'GREAT Britain' campaign: a government campaign aimed at promoting the best of the country abroad, encouraging people to visit, do business, invest and study in the country

ground rush: the sense of the ground rushing past you as you fly 100–300 feet above it

Hammerhead Break: a formation that involves pulling up for a Quarter Clover (a twist in the loop to ninety degrees), then banking back, pointing at the ground, at which point the planes split

Hawk-current: the term for a pilot who is confident and familiar with flying the Hawk T1 jet

Heart: a much-loved part of the Red Arrows display performed by the synchro pair

Heart and Spear: this involves two aircraft performing a vertical pull-up, then separating and rolling with the smoke on, to create the top part of the Heart shape, with Red 8 or 9 flying the Spear for the Heart

hot debrief: in-the-moment feedback

hypoxia: oxygen-deficiency

Infinity Break: an innovative and spectacular formation break in which Red 1 rolls around the smoke of the team to create the shape of the infinity symbol before the formation splits

JEngO: Junior Engineering Officer, who takes on the role of Circus 1 and flies in the back seat of Red 1's aircraft during the display season

kneeboard: a rectangular area (A5 paper size) attached the G-trousers on the top of the thigh area so that a pilot can write down any notes they need

Lancaster: a formation shape depicting a Lancaster Bomber

Leader's Benefit: a shape flown from the inception of the team: Reds 2 and 3 fly in an arrow position on the leader while 4 and 5 fly in line abreast

Line Abreast: a formation position with the aircraft flying side-by-side with the nose of every aircraft in alignment with each other

Line Astern: a formation position with the aircraft flying directly behind each other

MAA: the Military Aviation Authority, the regulating body of all defence aviation activities, including the Red Arrows team

metronomic cadence: the particular rhythm and intonation with which the leader's voice-commands are delivered to the other pilots

Mirror Flat / Mirror Roll: a manoeuvre performed by the synchro pair, together with Reds 8 and 9. As they fly in a vertical column, Reds 6 and 7 invert themselves so that they are directly above Reds 8 and 9, with the canopies of their jets just thirty feet apart

MOD: Ministry of Defence

nautical miles: Nautical miles are used to measure the distance travelled through water. A nautical mile is slightly longer than a mile on land, equalling 1.1508 land-measured (or statute) miles. The nautical mile is based on the Earth's longitude and latitude coordinates, with one nautical mile equalling one minute of latitude

Nav: the pilot given responsibility for planning a sortie; the role is shared between pilots during the season

nine-ship day: the 'first nine-ship day' is the first time Reds 1-9 practise that season's intended display together. Until now, it has been in sections of four, five or seven aircraft. For this reason, this moment is seen as the very first nine aircraft rehearsal

OC: the Officer Commanding the Red Arrows (Red 11)

'Open up and relax': thirty to sixty seconds of reprieve from aerobatic flying by remaining straight and level. This offers a chance for pilots to breathe, collect thoughts and allow the more experienced members of the formation to offer some debrief comments to the new pilots

opposition pass: a head-to-head pass between the synchro pair (Reds 6 and 7). They pass no closer than 100 feet from each other at a closure speed between 700 and 800 mph

out brief: the final check that everything is in order prior to flying

over-perform (the aircraft): to manoeuvre the aircraft above the maximum amount of lift available

Palm Split: a manoeuvre by all nine aircraft that depicts a palm tree

pannier: a two- by three-foot space for bags in each Hawk jet

PDA: Public Display Authority, the annual assessment of the Red Arrows

Phoenix: the widest manoeuvre of all the formation shapes flown during the first half of the display. From the right wing tip of Red 8 to the left wing-tip of Red 9, Phoenix spans 320 feet

pigz: the Red Arrows fine system, named after the 'Pigz Boards' found at most RAF squadrons. When a minor mistake or misdemeanour is made by an individual (or group), a small fine is given (e.g. 50p, £1). The Pigz Board will get filled up with the fines (and names of individuals), which are then announced at the end of the week

pulling G: an expression to denote the sensation of gravitational force (G-force) felt on the body due to an acceleration of the aircraft. More specifically, G-force is a measurement of the type of force per unit mass – typically acceleration – that causes a perception of weight, with a G-force of 1G equal to the conventional value of gravitational acceleration on Earth, g, of about 9.8 m/s2

Pyramid: a formation shape that depicts a triangular, pyramid shaped disposition

Python: a manoeuvre in which Enid flies two big barrel rolls, one after another, covering six miles of ground-track

QFI: Qualified Flying Instructor

QRA: Quick Reaction Alert duties

Quarter Clover: a manoeuvre that involves pulling up into a loop and twisting the formation by ninety degrees while pointing straight up, before pulling down to complete the vertical element of the man-oeuvre, perpendicular to where you began

RAFAT: Royal Air Force Aerobatic Team, the official name of the Red Arrows

Reds 1 to 9: the pilots of the Red Arrows, named after their red flying suits

Red 10: the Red Arrow whose primary role is to check on safety and act as display commentator on the ground

RIAT: Royal International Air Tattoo, the largest military air display in the world, now held at RAF Fairford in Gloucestershire

Rollback: a manoeuvre in which a pair of jets take turns to pull up out of formation, roll 360 degrees in a tight barrel roll, then slot back perfectly on to the outer edges of the formation

rolling display: this consists of rolls and wing-over manoeuvres and will be flown if the cloud base is clear up to 2,500 feet

Rolling Heart: an adaptation of the Heart formation, which can be performed as part of the rolling display

Roulette: a low-level 360-degree turning manoeuvre with the aircraft flying in opposite directions

RT: radio, short for radio transmitter

run-in: when pilots start to line up for the run-in track to the show, about four minutes and twenty miles from the site

safety box: the area in which pilots are never too far from or too close to the aircraft next to them

see-off: a 'formal start' performed by the Circus, in which the engineers' pre-flight procedures and checks are synchronised

SEngO: Senior Engineering Officer

Shakedown flight: when Circus members are taken up for the first time in a Hawk

smoke plot: planning to ensure no one runs out of smoke during a display

'Smoke on, Go!': the call Red 1 makes to tell the other pilots to turn their smoke trails on

SOPs: Standard Operating Procedures

sortie: French for 'military mission', defined as an operation carried out by a deployed unit, which can be an aircraft, ship or a group of people. We use it to mean each flight by one or more aircraft

Spaghetti Break: a spectacular move that sees the nine aircraft fan out from the top of a straight loop before landing

Spitfire: a formation shape that depicts the iconic Supermarine Spitfire

Springhawk: the final stage of pre-season training exercises abroad

STEM subjects: science, technology, engineering and maths

step-down process: flying at increasingly low heights, ranging from 3,000 feet down to 500 feet

stream take-off: multiple aircraft line up on the runway and take off individually at timed intervals up to five seconds apart

Swan: a graceful formation shape that has Reds 1, 6 and 7 flying in Line Astern with Reds 2, 3, 4, 5, 8 and 9 flying in a V-shape behind Red 7. It is one of the most demanding first half shapes to fly

synchro leader: Red 6 – the more experienced pilot in the synchro pair, usually in their third (and often final) year with the team, having flown as Red 7 during the previous season. Heading into their third year to fly in the Red 6 position, this pilot will have chosen a team member (from the cadre of first year pilots, all with one season just

completed) to become Red 7 in their second year and train intensively together to perform the various synchro-specific manoeuvres

synchro pair: Reds 6 and 7, who make up the rear section of the formation

tattoo: originally refers to a form of military music, but has since evolved
to be generally applied to more elaborate military shows and exhibi-
tions, such as the Royal International Air Tattoo. The military term
is unrelated to the Tahitian origins of an ink tattoo

tip-in: when preparing to land, the pilot puts the wheels and flaps of the
jet into the correct configuration for landing. The jet is pointing 180
degrees from the runway (called being 'downwind'), so that when the
pilot is ready to turn, they bank the aircraft to about 60 degrees (called
the 'tip-in') to turn the aircraft vector towards the landing runway

Tornado: a manoeuvre in which the front seven aircraft fly in an arrow
position while Reds 8 and 9 perform very tight barrel rolls, so that
they 'hug' the trailing smoke of the seven aircraft in front

Tornado F3: the Air Defence Variant of the Panavia Tornado, developed
by the United Kingdom as a dedicated long-range interceptor fighter

transit take-off: a formation take-off used for up to eleven aircraft when
transiting from one location to another

Twinkle Roll: a roll conducted by a formation of aircraft as an aerobatic
manoeuvre in which every aircraft does a full 360-degree revolution

Twizzle: a formation in which the aircraft fly so that they are lined up
on each other's left wing and in close succession, rolling to the right
through 270 degrees and under those following

under load: maintaining G-force via acceleration, vertical manoeuvring
and sharp turns; this helps pilots to keep their position within a formation, resulting in tight formations without any quivering wings

Vertical Break: the aircraft loop upwards then break while pointing vertically down at the ground, producing a big fan of smoke as they separate

Vice-Versa: a manoeuvre that sees Red 6 flying inverted as the pilot heads towards Red 7, the pair rotating as they pass each other so that Red 7 becomes the one flying inverted, with Red 6 now the right way up

Vixen Break: in this manoeuvre Enid is with Reds 8 and 9 in Line Abreast, flying in the direction of the crowd before they all fan out with a perfect plume of red, white and blue smoke

Wall: a manoeuvre in which the jets arrive spaced out really widely in a line abreast

WHAM: stands for What Happens According to Mange, the complex planning document that features a minute-by-minute schedule with which check-ins and events must correlate during the display season. Mange is an affectionate term for 'Manager', used extensively in the past by the Red Arrows – with the Team Manager having previously been the person who coordinated and oversaw this detailed plan

wheels: in RAF parlance, this relates to any bike, car, truck, tuk-tuk or coach that carries the pilots from one place to another

Whole Force: the RAF aviators (both regular personnel and full-time reserve colleagues), civil servants and a small number of contractors who make up the Blues

wingmen: other pilots in the formation, not including the flight leader

XO: the Executive Officer of the Red Arrows

For a full list of Red Arrows formations, including descriptions of how they're formed, see https://www.raf.mod.uk/display-teams/red-arrows/displays/

Roll of Honour

THE ROYAL AIR FORCE AEROBATIC TEAM

1965 Leader Flt Lt L Jones, 2 Flt Lt B A Nice, 3 Flt Lt R G Hanna, 4 Flt Lt G L Ranscombe, 5 Fg Off P G Hay, 6 Flt Lt R E W Loverseed, 7 Flt Lt H J D Prince, 8 Flt Lt E C F Tilsley, Manager Sqn Ldr R A E Storer, Engineer Fg Off D Green, Engineer Fg Off C T Harrow, Engineer Fg Off D Whitby

1966 Leader Sqn Ldr R G Hanna, 2 Flt Lt D A Bell, 3 Flt Lt R W Langworthy, 4 Flt Lt P R Evans, 5 Flt Lt R Booth, 6 Flt Lt H J D Prince, 7 Flt Lt T J G Nelson, 8 Flt Lt F J Hoare, 9 Flt Lt D McGregor, Manager Sqn Ldr R A E Storer, Engineer Fg Off C T Harrow, Engineer Fg Off D Whitby

1967 Leader Sqn Ldr R G Hanna, 2 Flt Lt D A Bell, 3 Flt Lt F J Hoare, 4 Flt Lt P R Evans, 5 Flt Lt R Booth, 6 Flt Lt H J D Prince, 7 Flt Lt E E Jones, Manager Flt Lt L G Wilcox, Engineer Fg Off D Whitby, Adjutant Flt Lt R Dench

1968 Leader Sqn Ldr R G Hanna, 2 Flt Lt D A Bell, 3 Flt Lt F J Hoare, 4 Flt Lt P R Evans, 5 Flt Lt I C H Dick, 6 Flt Lt R Booth, 7 Flt Lt J T Kingsley, 8 Flt Lt D A Smith, 9 Flt Lt R B Duckett, Manager Flt Lt L G Wilcox, Engineer Fg Off D Whitby, Adjutant Flt Lt R Dench

1969 Leader Sqn Ldr R G Hanna, 2 Flt Lt P R Evans, 3 Flt Lt D A Smith, 4 Flt Lt R B Duckett, 5 Flt Lt R Perreaux, 6 Flt Lt J T Kingsley, 7 Flt Lt I C H Dick, 8 Flt Lt J D Rust, 9 Sqn Ldr R P Dunn, Manager Flt Lt P Mackintosh, Engineer Fg Off G E White, Adjutant Flt Lt R Dench

1970 Leader Sqn Ldr D Hazell, 2 Flt Lt R Perreaux, 3 Flt Lt D A Smith, 4 Flt Lt J D Rust, 5 Flt Lt J Haddock, 6 Flt Lt I C H Dick, 7 Flt Lt R B Duckett, 8 Flt Lt D S B Marr, 9 Flt Lt R E W Loverseed, Manager Flt Lt P Mackintosh, Engineer Flt Lt G E White, Adjutant WO L Ludlow

1971 Leader Sqn Ldr R E W Loverseed, 2 Sqn Ldr D S B Marr, 3 Flt Lt A C East, 4 Flt Lt W B Aspinall, 5 Flt Lt P J J Day, 6 Flt Lt C F Roberts, 7 Flt Lt R E Somerville, Manager Flt Lt K J Tait, Engineer Flt Lt G E White, Adjutant WO L Ludlow

1972 Leader Sqn Ldr I C H Dick, 2 Flt Lt W B Aspinall, 3 Flt Lt A C East, 4 Flt Lt R E Somerville, 5 Flt Lt K J Tait, 6 Flt Lt P J J Day, 7 Flt Lt D Binnie, 8 Flt Lt E E G Girdler, 9 Flt Lt C F Roberts, Manager Flt Lt B Donnelly, Engineer Flt Lt I Brackenbury, Adjutant WO S Wild

1973 Leader Sqn Ldr I C H Dick, 2 Sqn Ldr W B Aspinall, 3 Flt Lt B Donnelly, 4 Flt Lt E E G Girdler, 5 Flt Lt K J Tait, 6 Flt Lt D Binnie, 7 Sqn Ldr R E Somerville, 8 Flt Lt D J Sheen, 9 Flt Lt P J J Day, Manager Flt Lt R M Joy, Engineer Flt Lt I Brackenbury, Adjutant WO H E D Runsdstrom

1974 Leader Sqn Ldr I C H Dick, 2 Flt Lt K J Tait, 3 Flt Lt B Donnelly, 4 Flt Lt E E G Girdler, 5 Flt Lt C M Phillips, 6 Flt Lt D Binnie, 7 Sqn Ldr R E Somerville, 8 Flt Lt D J Sheen, 9 Flt Lt R Eccles, Manager Flt Lt R M Joy, Engineer Flt Lt I Brackenbury, Adjutant WO H E D Runsdstrom

1975 Leader Sqn Ldr R B Duckett, 2 Flt Lt M J Phillips, 3 Flt Lt B Donnelly, 4 Flt Lt R Eccles, 5 Flt Lt J Blackwell, 6 Flt Lt D Sheen, 7 Sqn Ldr B R Hoskins, 8 Flt Lt M Cornwell, 9 Flt Lt R Barber, Manager Sqn Ldr A L Wall, Engineer Flt Lt A Hunt, Adjutant WO H E D Runsdstrom

1976 Leader Sqn Ldr R B Duckett, 2 Flt Lt M J Phillips, 3 Flt Lt R Eccles, 4 Flt Lt D R Carvell, 5 Flt Lt R S Barber, 6 Sqn Ldr B R Hoskins, 7 Flt Lt M Cornwell, 8 Flt Lt M T Curley, 9 Flt Lt N S Champness, Manager Sqn Ldr A L Wall, Engineer Flt Lt A Hunt, Adjutant WO H G Thorne

1977 Leader Sqn Ldr F J Hoare, 2 Flt Lt D R Carvell, 3 Flt Lt R S Barber, 4 Flt Lt M J Phillips, 5 Flt Lt N S Champness, 6 Flt Lt M Cornwell, 7 Flt Lt M T Curley, 8 Flt Lt R M Thomas, 9 Flt Lt M B Stoner, Manager Flt Lt M B Whitehouse, Engineer Flt Lt A Hunt, Adjutant WO H G Thorne

1978 Leader Sqn Ldr F J Hoare, 2 Flt Lt D R Carvell, 3 Flt Lt M B Stoner, 4 Flt Lt M J Phillips, 5 Flt Lt L A Grose, 6 Flt Lt M T Curley, 7 Flt Lt R M Thomas, 8 Flt Lt S R Johnson, 9 Flt Lt B C Scott, Manager Flt Lt M B Whitehouse, Engineer Flt Lt R A Lewis, Adjutant WO H G Thorne

1979 Leader Sqn Ldr B R Hoskins, 2 Flt Lt M T Curley, 3 Flt Lt B C Scott, 4 Flt Lt M D Howell, 5 Flt Lt M B Stoner, 6 Flt Lt R M Thomas, 7 Sqn Ldr S R Johnson, 8 Flt Lt N J Wharton, 9 Flt Lt W Ward, Manager Sqn Ldr R Thilthorpe, Engineer Flt Lt R A Lewis, Adjutant WO H G Thorne

1980 Leader Sqn Ldr B R Hoskins, 2 Flt Lt M D Howell, 3 Flt Lt W Ward, 4 Flt Lt N J Wharton, 5 Flt Lt B C Scott, 6 Flt Lt R M Thomas, 7 Sqn Ldr S R Johnson, 8 Flt Lt B S Walters, 9 Flt Lt T R Watts, Manager Sqn Ldr R Thilthorpe, Engineer Flt Lt R A Lewis, Adjutant WO H G Thorne

1981 Leader Sqn Ldr B R Hoskins, 2 Flt Lt B F Walters, 3 Flt Lt W Ward, 4 Flt Lt M H de Courcier, 5 Flt Lt N J Wharton, 6 Sqn Ldr S R Johnson, 7 Flt Lt T R Watts, 8 Flt Lt I J Huzzard, 9 Flt Lt J R Myers, Manager Sqn Ldr R Thilthorpe, Engineer Flt Lt G M Nisbet, Adjutant WO H G Thorne

1982 Leader Sqn Ldr J Blackwell, 2 Flt Lt B S Walters, 3 Flt Lt J R Myers, 4 Flt Lt I J Huzzard, 5 Flt Lt W Ward, 6 Flt Lt T R Watts, 7 Flt Lt M H de Courcier, 8 Flt Lt T W L Miller, 9 Flt Lt P A Tolman, Manager Sqn Ldr R Thilthorpe, Engineer Flt Lt G M Nisbet, Adjutant WO H G Thorne

1983 Leader Sqn Ldr J Blackwell, 2 Sqn Ldr I J Huzzard, 3 Flt Lt J R Myers, 4 Flt Lt T W L Miller, 5 Flt Lt E H Ball, 6 Flt Lt M H de Courcier, 7 Flt Lt P A Tolman, 8 Flt Lt S H Bedford, 9 Flt Lt C A R Hirst, Manager Sqn Ldr J E Steenson, Engineer Flt Lt M E J Render, Adjutant WO H G Thorne

1984 Leader Sqn Ldr J Blackwell, 2 Flt Lt S H Bedford, 3 Flt Lt G I Hannam, 4 Sqn Ldr T W L Miller, 5 Sqn Ldr E H Ball, 6 Flt Lt P A Tolman, 7 Flt Lt A R Boyens, 8 Flt Lt P D Lees, 9 Flt Lt A K Lunnon-Wood, Manager Sqn Ldr J E Steenson, Engineer Flt Lt M E J Render, Adjutant WO D H A Chubb

1985 Leader Sqn Ldr R M Thomas, 2 Flt Lt P D Lees, 3 Sqn Ldr E H Ball, 4 Flt Lt S H Bedford, 5 Sqn Ldr G I Hannam, 6 Flt Lt A R Boyens, 7 Flt Lt A K Lunnon-Wood, 8 Flt Lt C D R McIlroy, 9 Sqn Ldr A B Chubb, Manager Sqn Ldr H R Ploszek, Engineer Flt Lt M E J Render, Adjutant WO D H A Chubb

1986 Leader Sqn Ldr R M Thomas, 2 Flt Lt P D Lees, 3 Sqn Ldr A B Chubb, 4 Flt Lt P J Collins, 5 Sqn Ldr G I Hannam, 6 Flt Lt A K Lunnon-Wood, 7 Flt Lt C D R McIlroy, 8 Flt Lt D W Findlay, 9 Flt Lt A P Thurley, Manager Sqn Ldr H R Ploszek, Engineer Flt Lt J S Chantry, Adjutant WO D H A Chubb

1987 Leader Sqn Ldr R M Thomas, 2 Sqn Ldr P J Collins, 3 Flt Lt M A Carter, 4 Flt Lt M J Newbery, 5 Sqn Ldr A B Chubb, 6 Flt Lt C D R McIlroy, 7 Flt Lt A P Thurley, 8 Flt Lt J E Rands, 9 Flt Lt G M Bancroft-Wilson, Manager Sqn Ldr H R Ploszek, Engineer Flt Lt J S Chantry, Adjutant WO M R J Fleckney

1988 Leader Sqn Ldr T W L Miller, 2 Flt Lt G M Bancroft-Wilson, 3 Flt Lt D C Riley, 4 Sqn Ldr P J Collins, 5 Flt Lt S W M Johnson, 6 Sqn Ldr A P Thurley, 7 Flt Lt J E Rands, 8 Sqn Ldr J W Glover, 9 Flt Lt M A Carter, Manager Sqn Ldr H R Ploszek, Engineer Flt Lt J D Williams, Adjutant WO M R J Fleckney

1989 Leader Sqn Ldr T W L Miller, 2 Flt Lt A W Hoy, 3 Flt Lt M J H Cliff, 4 Flt Lt G M Bancroft-Wilson, 5 Sqn Ldr D C Riley, 6 Flt Lt J E Rands, 7 Flt Lt S W M Johnson, 8 Flt Lt J W Glover, 9 Flt Lt M J M Newton, Manager Sqn Ldr A J Stewart, Engineer Flt Lt J D Williams, Adjutant WO M R J Fleckney

1990 Leader Sqn Ldr T W L Miller, 2 Flt Lt A Smith, 3 Flt Lt P C H Rogers, 4 Flt Lt A W Hoy, 5 Sqn Ldr D C Riley, 6 Flt Lt S W M Johnson, 7 Flt Lt J M Newton, 8 Flt Lt D A Wyatt, 9 Flt Lt M J H Cliff, Manager Sqn Ldr A J Stewart, Engineer Flt Lt C R Bushell, Adjutant WO M R J Fleckney

1991 Leader Sqn Ldr A P Thurley, 2 Flt Lt G P Howes, 3 Flt Lt N C Rogers, 4 Flt Lt A Smith, 5 Flt Lt S C Meade, 6 Flt Lt J M Newton, 7 Flt Lt D A Wyatt, 8 Flt Lt A W Hoy, 9 Flt Lt M J H Cliff, Manager Sqn Ldr A J Stewart, Engineer Flt Lt C R Bushell, Adjutant WO M R J Fleckney

1992 Leader Sqn Ldr A P Thurley, 2 Flt Lt R W Last, 3 Flt Lt B J Cross, 4 Sqn Ldr G P Howes, 5 Flt Lt J C Bird, 6 Sqn Ldr D A Wyatt, 7 Flt Lt S C Meade, 8 Flt Lt A Smith, 9 Flt Lt N C Rogers, Manager Sqn Ldr L Garside-Beattie, Engineer Flt Lt R L Miller, Adjutant WO M R J Fleckney

1993 Leader Sqn Ldr A P Thurley, 2 Flt Lt S Chiddention, 3 Flt Lt M G Ball, 4 Flt Lt B J Cross, 5 Flt Lt J C Bird, 6 Sqn Ldr S C Meade, 7 Flt Lt R W Last, 8 Sqn Ldr G P Howes, 9 Flt Lt N C Rogers, Manager Sqn Ldr L Garside-Beattie, Engineer Flt Lt R L Miller, Adjutant WO J Howard

1994 Leader Sqn Ldr J E Rands, 2 Flt Lt C D Jepson, 3 Flt Lt M W Zanker, 4 Flt Lt K P Truss, 5 Flt Lt M G Ball, 6 Flt Lt R W Last, 7 Flt Lt S Chiddention, 8 Flt Lt B J Cross, 9 Flt Lt J C Bird, Manager Sqn Ldr L Garside-Beattie, Engineer Flt Lt M J Northover, Adjutant WO J Howard

1995 Leader Sqn Ldr J E Rands, 2 Flt Lt R Matthews, 3 Flt Lt S D Perrett, 4 Flt Lt T Couston, 5 Flt Lt M W Zanker, 6 Flt Lt S Chiddention, 7 Sqn Ldr K P Truss, 8 Flt Lt C D Jepson, 9 Sqn Ldr M G Ball, Manager Sqn Ldr H M Williams, Engineer Flt Lt M J Northover, Adjutant WO J Howard

1996 Leader Sqn Ldr J E Rands, 2 Sqn Ldr A C Offer, 3 Flt Lt D N Stobie, 4 Flt Lt R Matthews, 5 Flt Lt S D Perrett, 6 Sqn Ldr K P Truss, 7 Flt Lt T Couston, 8 Flt Lt C D Jepson, 9 Flt Lt M W Zanker, Manager Sqn Ldr H M Williams, Engineer Flt Lt M J Northover, Adjutant WO J Howard

1997 Leader Sqn Ldr S C Meade, 2 Flt Lt I S Smith, 3 Flt Lt G M Water-fall, 4 Sqn Ldr A C Offer, 5 Flt Lt A Cubin, 6 Flt Lt T Couston, 7 Flt Lt D N Stobie, 8 Flt Lt R Matthews, 9 Flt Lt S D Perrett, Manager Sqn Ldr H M Williams, SEngO Flt Lt D Chowns, JEngO Flt Lt J Russell, Adjutant WO J Howard

1998 Leader Sqn Ldr S C Meade, 2 Flt Lt A D E Evans, 3 Flt Lt K A Lewis, 4 Flt Lt I S Smith, 5 Flt Lt R P G Patounas, 6 Flt Lt D N Stobie, 7 Flt Lt A Cubin, 8 Sqn Ldr A C Offer, 9 Sqn Ldr G M Waterfall, 10 Flt Lt R R Jones, Manager Sqn Ldr E E Webster, SEngO Flt Lt D Chowns, JEngO Flt Lt J Russell, Adjutant WO J Howard

1999 Leader Wg Cdr S C Meade, 2 Flt Lt J D Provost, 3 Flt Lt M R Cut-more, 4 Flt Lt A D E Evans, 5 Sqn Ldr K A Lewis, 6 Sqn Ldr A Cubin, 7 Flt Lt R P G Patounas, 8 Flt Lt I S Smith, 9 Sqn Ldr G M Waterfall, 10 Flt Lt R R Jones, Manager Sqn Ldr J M Paige, SEngO Flt Lt G Mar-tin, JEngO Flt Lt A D McNeill, Adjutant WO J Howard

2000 Leader Sqn Ldr A C Offer, 2 Flt Lt J R Hawker, 3 Flt Lt J P Hughes, 4 Flt Lt C D Carder, 5 Flt Lt M R Cutmore, 6 Flt Lt R P G Patounas, 7 Flt Lt J D Provost, 8 Flt Lt A D E Evans, 9 Sqn Ldr K A Lewis, 10 Flt Lt R R Jones, Manager Sqn Ldr J M Paige, SEngO Sqn Ldr M J Nor-thover, JEngO Flt Lt A D McNeill, Adjutant WO J H May

2001 Leader Wg Cdr A C Offer, 2 Flt Lt A F Parkinson, 3 Flt Lt C Gleave, 4 Sqn Ldr M M Garland, 5 Flt Lt J P Hughes, 6 Sqn Ldr J D Provost, 7 Flt Lt J R Hawker, 8 Sqn Ldr C D Carder, 9 Flt Lt M R Cutmore, 10 Sqn Ldr A D E Evans, Manager Sqn Ldr J M Paige, SEngO Sqn Ldr M J Northover, JEngO Flt Lt T Beagle, Adjutant WO J H May

2002 Leader Sqn Ldr C D Jepson, 2 Flt Lt D Thomas, 3 Sqn Ldr J H Green, 4 Flt Lt A F Parkinson, 5 Sqn Ldr C Gleave, 6 Sqn Ldr J R Hawker, 7 Sqn Ldr M M Garland, 8 Sqn Ldr C D Carder, 9 Flt Lt J P Hughes, 10 Flt Lt S C Underwood, Manager Sqn Ldr L C Johnson, SEngO Sqn Ldr M J Northover, JEngO Flt Lt T Beagle, Adjutant WO J H May

2003 Leader Sqn Ldr C D Jepson, 2 Flt Lt J P Griggs, 3 Flt Lt D C Mason, 4 Flt Lt D J Simmons, 5 Sqn Ldr J H Green, 6 Sqn Ldr M M Garland, 7 Sqn Ldr D Thomas, 8 Flt Lt A F Parkinson, 9 Sqn Ldr C Gleave, 10 Flt Lt S C Underwood, Manager Sqn Ldr S E Varley, PRO Miss

R L Huxford, SEngO Sqn Ldr R K Carleton, JEngO Flt Lt T Beagle, Adjutant WO J H May

2004 Leader Sqn Ldr C D Jepson, 2 Flt Lt A F Parkinson, 3 Flt Lt S D Stevens, 4 Flt Lt D J Slow, 5 Sqn Ldr D C Mason, 6 Sqn Ldr D Thomas, 7 Flt Lt D J Simmons, 8 Flt Lt J P Griggs, 9 Sqn Ldr J H Green, 10 Flt Lt S C Underwood, Manager Sqn Ldr S E Varley, PRO Miss R L Huxford, SEngO Sqn Ldr R K Carleton, JEngO Flt Lt S C Race, Adjutant WO J H May

2005 Leader Sqn Ldr R P G Patounas, 2 Flt Lt S Morley, 3 Flt Lt M J Higgins, 4 Flt Lt D J Slow, 5 Flt Lt J H Turner, 6 Sqn Ldr D J Simmons, 7 Flt Lt S D Stevens, 8 Flt Lt J P Griggs, 9 Sqn Ldr D C Mason, 10 Flt Lt S C Underwood, Manager Sqn Ldr S E Varley, PRO Miss R L Huxford, SEngO Sqn Ldr S R Davies, JEngO Flt Lt S C Race, Adjutant WO J H May

2006 Leader Wg Cdr R P G Patounas, 2 Flt Lt G B J Perilleux, 3 Flt Lt D R Ellacott, 4 Sqn Ldr S Morley, 5 Flt Lt P O'Grady, 6 Flt Lt S D Stevens, 7 Flt Lt J H Turner, 8 Flt Lt D J Slow, 9 Sqn Ldr M J Higgins, 10 Flt Lt A C R Robins, Manager Sqn Ldr P J Hunt, PRO Miss R L Huxford, SEngO Sqn Ldr S R Davies, JEngO Flt Lt R D J Gates, Adjutant WO J H May

2007 Leader Wg Cdr J R Hawker, 2 Sqn Ldr B D Murphy, 3 Flt Lt A R Keith, 4 Flt Lt G B J Perilleux, 5 Flt Lt D R Ellacott, 6 Sqn Ldr J H Turner, 7 Flt Lt P O'Grady, 8 Sqn Ldr S Morley, 9 Sqn Ldr M J Higgins, 10 Flt Lt A C R Robins, Manager Sqn Ldr P J Hunt, PRM Miss R L Huxford, SEngO Sqn Ldr E D Williams, JEngO Flt Lt A Scott, Adjutant WO J H May

2008 Leader Wg Cdr J R Hawker, 2 Flt Lt S P Rea, 3 Flt Lt M R Ling, 4 Sqn Ldr G Duff, 5 Flt Lt A R Keith, 6 Flt Lt P O'Grady, 7 Sqn Ldr B D Murphy, 8 Flt Lt G B J Perilleux, 9 Flt Lt D R Ellacott, 10 Flt Lt A C R Robins, Manager Sqn Ldr J S Trott, PRM Miss R L Huxford, PRO Miss E J Thomas, SEngO Sqn Ldr E D Williams, JEngO Flt Lt C R Fenn, Adjutant WO J H May

2009 Leader Wg Cdr J R Hawker, 2 Flt Lt Z R Sennett, 3 Flt Lt D A Montenegro, 4 Flt Lt D B Davies, 5 Flt Lt S P Rea, 6 Sqn Ldr B D Murphy, 7 Flt Lt M R Ling, 8 Sqn Ldr G Duff, 9 Flt Lt A R Keith, 10 Sqn Ldr G Bagnall, Manager Sqn Ldr J S Trott, PRM Miss N L Wright, PRO Miss E J Thomas, SEngO Sqn Ldr G P Ball, JEngO Flt Lt C Fenn, Adjutant WO A Murray BEM

2010 Leader Sqn Ldr B D Murphy, 2 Flt Lt B M Plank, 3 Flt Lt K A Moore, 4 Flt Lt D B Davies, 5 Flt Lt Z R Sennett, 6 Flt Lt M R Ling/ Flt Lt P O'Grady, 7 Flt Lt D A Montenegro, 8 Sqn Ldr G Duff, 9 Flt Lt S P Rea, 10 Sqn Ldr G Bagnall, Manager Sqn Ldr J S Trott/ Sqn Ldr E J Parker, PRM Miss N L Wright/ Miss E J Thomas, SEngO Sqn Ldr G P Ball, JEngO Flt Lt A Bryant, Adjutant WO A Murray BEM

2011 Leader Sqn Ldr B D Murphy, 2 Flt Lt C Lyndon-Smith, 3 Flt Lt S Cunningham, 4 Flt Lt J W J Egging, 5 Flt Lt K A Stewart, 6 Flt Lt D A Montenegro, 7 Flt Lt B M Plank, 8 Flt Lt D B Davies, 9 Flt Lt Z R Sennett, 10 Sqn Ldr G Bagnall, Manager Sqn Ldr E J Parker, PRM Miss E J Thomas, SEngO Sqn Ldr R Priday, JEngO Flt Lt A B Littler, Adjutant WO A Murray BEM

2012 Leader Sqn Ldr J Turner, 2 Flt Lt M Pert, 3 Flt Lt M Child, 4 Flt Lt J McMillan, 5 Sqn Ldr M Higgins, 6 Flt Lt B M Plank, 7 Flt Lt C Lyndon-Smith, 8 Flt Lt D B Davies, 10 Sqn Ldr M Ling, Manager Sqn Ldr E J Parker, PRM Mrs J Cross, SEngO Sqn Ldr R Priday, JEngO Flt Lt A B Littler, Adjutant WO A Murray BEM

2013 Leader Sqn Ldr J Turner, 2 Flt Lt O Parr, 3 Flt Lt M Lawson, 4 Flt Lt M Pert, 5 Flt Lt S Morris, 6 Flt Lt C Lyndon-Smith, 7 Flt Lt J McMillan, 8 Flt Lt B M Plank, 9 Flt Lt M Child, 10 Sqn Ldr M Ling, Manager Sqn Ldr R Shackleton, PRM Mr A Morton, PRO Mrs J Pearson, SEngO Sqn Ldr J Fortune, JEngO Flt Lt A B Littler, Adjutant WO A Irons

2014 Leader Sqn Ldr J Turner, 2 Flt Lt S Campbell, 3 Flt Lt J Hourston, 4 Flt Lt O Parr, 5 Flt Lt S Morris, 6 Flt Lt J McMillan, 7 Flt Lt M Lawson, 8 Flt Lt M Pert, 9 Flt Lt M Child, 10 Sqn Ldr M Ling, Manager Sqn Ldr R Shackleton, PRM Mr A Morton, PRO Mrs J Pearson, SEngO Sqn Ldr J Fortune/ Sqn Ldr H Raja, JEngO Flt Lt M Noye, Adjutant WO A Irons

2015 OC RAFAT Wg Cdr M Higgins, Leader Sqn Ldr D Montenegro, 2 Flt Lt M Bowden, 3 Flt Lt E Cox, 4 Flt Lt S Campbell, 5 Flt Lt T Bould, 6 Flt Lt M Lawson, 7 Flt Lt S Morris, 8 Flt Lt O Parr, 9 Flt Lt J Hourston, 10 Sqn Ldr M Ling, Manager Sqn Ldr C Driscoll, PRM Mr A Morton, PRO Mrs J Pearson, SEngO Sqn Ldr P Searle, JEngO Flt Lt M Ramsden, Adjutant WO A Irons

2016 OC RAFAT Wg Cdr M Higgins, Leader Sqn Ldr D Montenegro, 2 Flt Lt M Masters, 3 Flt Lt S Taylor, 4 Flt Lt M Bowden, 5 Flt Lt E Cox, 6 Flt Lt S Morris, 7 Flt Lt T Bould, 8 Flt Lt S Campbell, 9 Flt

Lt J Hourston, 10 Sqn Ldr M Ling, Manager Sqn Ldr C Driscoll, PRM Mr A Morton, PRO Mrs J Pearson, APO Sqn Ldr D Platt, SEngO Sqn Ldr Pete Searle, JEngO Flt Lt M Ramsden, Adjutant WO A Irons

2017 OC RAFAT Wg Cdr M Higgins/ A Keith, Leader Sqn Ldr D Montenegro, 2 Flt Lt T Keeley, 3 Flt Lt Dan Lowes, 4 Flt Lt M Masters, 5 Flt Lt C Lyndon-Smith, 6 Flt Lt T Bould, 7 Flt Lt S Taylor, 8 Flt Lt M Bowden, 9 Flt Lt E Cox, 10 Sqn Ldr M Ling, Manager Sqn Ldr C Driscoll, PRM Mr A Morton, PRO Mrs J Pearson, APO Sqn Ldr D Platt, SEngO Sqn Ldr R Bland, JEngO Flt Lt A Mason, Adjutant WO A Irons

2018 OC RAFAT Wg Cdr A Keith, Leader Sqn Ldr M Pert, 2 Flt Lt J Bond, 3 Sqn Ldr M Ling, 4 Flt Lt C Lyndon-Smith, 5 Flt Lt D Lowes, 6 Flt Lt S Taylor, 7 Flt Lt T Keeley, 8 Flt Lt M Masters, 9 Flt Lt M Bowden, 10 Sqn Ldr A Collins, Manager Sqn Ldr D Smith, PRM Mr A Morton, PRO Fg Off M Fox, APO Flt Lt D Stark, SEngO Sqn Ldr R Bland, JEngO Flt Lt A Mason, Adjutant WO A Irons

2019 OC RAFAT Wg Cdr A Keith, Leader Sqn Ldr M Pert, 2 Flt Lt D Green, 3 Sqn Ldr M Bowden, 4 Flt Lt G Ogston, 5 Sqn Ldr S Morris, 6 Flt Lt T Keeley, 7 Flt Lt J Bond, 8 Flt Lt C Lyndon-Smith, 9 Flt Lt D Lowes, 10 Sqn Ldr A Collins, Manager Sqn Ldr D Smith, PRM Mr A Morton, PRO Fg Off M Fox, APO Sqn Ldr D Platt, SEngO Sqn Ldr R Bland, JEngO Flt Lt A Mason, Adjutant WO A Irons

2020 OC RAFAT Wg Cdr D Montenegro, Leader Sqn Ldr M Pert, 2 Flt Lt W Cambridge, 3 Flt Lt N Critchell, 4 Flt Lt J Turner, 5 Flt Lt D Simmonds, 6 Flt Lt J Bond, 7 Sqn Ldr G Ogston, 8 Flt Lt D Green, 9 Sqn Ldr S Morris, 10 Sqn Ldr A Collins, Manager Sqn Ldr D Smith, PRM Mr A Morton, PRO Fg Off V Delaney, APO Sqn Ldr D Platt, SEngO Sqn Ldr C Newcombe, JEngO Flt Lt B Ireland, Adjutant WO A Irons

2021 OC RAFAT Wg Cdr D Montenegro, Leader Sqn Ldr T Bould, 2 Flt Lt W Cambridge, 3 Sqn Ldr N Critchell, 4 Flt Lt J Turner, 5 Flt Lt D Simmonds, 6 Sqn Ldr J Bond, 7 Sqn Ldr G Ogston, 8 Flt Lt D Green, 9 Sqn Ldr S Morris, 10 Sqn Ldr A Collins, Manager Sqn Ldr D Smith, PRM Mr A Morton, PRO Fg Off V Delaney, APO Sqn Ldr D Platt, SEngO Sqn Ldr C Newcombe, JEngO Flt Lt J Fallon, Adjutant WO A Irons

2022 OC RAFAT Wg Cdr D Montenegro, Leader Sqn Ldr T Bould, 2 Flt Lt S Roberts, 3 Flt Lt P Kershaw, 4 Flt Lt W Cambridge, 5 Sqn Ldr N Critchell, 6 Sqn Ldr G Ogston, 7 Flt Lt J Turner, 8 Sqn Ldr J Bond, 9 Flt Lt D Simmonds, 10 Sqn Ldr G Muscat, Manager Sqn Ldr D Smith, PRM Mr A Morton, PRO Fg Off C Tavares-McKoy, SEngO Sqn Ldr C Phipps, JEngO Flt Lt A Addison, Adjutant WO A Irons

Bibliography

BOOKS

Callaway, Tim, *The Red Arrows – Celebrating 50 Display Seasons* (Morton Media Group, 2014)

Grant, R. G., *Flight: 100 Years of Aviation* (Dorling Kindersley, 2002)

Higgs, Colin, *Story of the Red Arrows* (Demand Media Limited, 2014)

March, Peter R., *50 Years of the Red Arrows* (The History Press, 2014)

Napier, Michael, *Tornado over the Tigris* (Pen and Sword Aviation, 2015)

Miller, T., Hanna, R. and Gibson, A., *25 Years of the Red Arrows* (Stanley Paul & Co, 1990)

Vanhoenacker, Mark, *How to Land a Plane* (Quercus, 2017)

Vanhoenacker, Mark, *Skyfaring: A Journey with a Pilot* (Vintage, 2015)

Watkins, David, *The History of RAF Aerobatic Teams from 1920* (Pen & Sword, 2010)

WEBSITES

https://historicengland.org.uk/images-books/publications/raf-scampton-historic-characterisation/raf-scampton-historic-characterisation/

https://www.raf.mod.uk

https://www.thetimes.co.uk/article/todays-air-pageant-at-hendon-2nd3x3f6k

https://www.nelsam.org.uk/NEAR/Events/Empire/UsworthEmpire39.htm

http://www.centralflyingschool.org.uk/History/History8.htm

https://www.docdroid.com/teeQIIm/royal-air-force-yearbook-1996-pdf#page=12

https://www.theguardian.com/artanddesign/2012/sep/17/strachan-stockwell-humber-bridge

PODCAST

North American Tour 2019 (BBC)

Index